T0234831

Statistical Thermodynamics
of Surfaces, Interfaces,
and Membranes

This book is dedicated to the memory of
Professor Alex Silberberg,
a pioneer in the field of complex fluids

Statistical Thermodynamics of Surfaces, Interfaces, and Membranes

Samuel A. Safran

Weizmann Institute of Science
Rehovot, Israel

Advanced Book Program

CRC Press is an imprint of the
Taylor & Francis Group, an **informa** business

First published 2003 by Westview Press

Published 2018 by CRC Press
Taylor & Francis Group
6000 Broken Sound Parkway NW, Suite 300
Boca Raton, FL 33487-2742

CRC Press is an imprint of the Taylor & Francis Group, an informa business

Copyright © 2003 by Taylor & Francis Group LLC

No claim to original U.S. Government works

This book contains information obtained from authentic and highly regarded sources. Reason-able efforts have been made to publish reliable data and information, but the author and publisher cannot assume responsibility for the validity of all materials or the consequences of their use. The authors and publishers have attempted to trace the copyright holders of all material reproduced in this publication and apologize to copyright holders if permission to publish in this form has not been obtained. If any copyright material has not been acknowledged please write and let us know so we may rectify in any future reprint.

Except as permitted under U.S. Copyright Law, no part of this book may be reprinted, reproduced, transmitted, or utilized in any form by any electronic, mechanical, or other means, now known or hereafter invented, including photocopying, microfilming, and recording, or in any information storage or retrieval system, without written permission from the publishers.

For permission to photocopy or use material electronically from this work, please access www. copyright.com (http://www.copyright.com/) or contact the Copyright Clearance Center, Inc. (CCC), 222 Rosewood Drive, Danvers, MA 01923, 978-750-8400. CCC is a not-for-profit organiza-tion that provides licenses and registration for a variety of users. For organizations that have been granted a photocopy license by the CCC, a separate system of payment has been arranged.

Trademark Notice: Product or corporate names may be trademarks or registered trademarks, and are used only for identification and explanation without intent to infringe.

Visit the Taylor & Francis Web site at
http://www.taylorandfrancis.com

and the CRC Press Web site at
http://www.crcpress.com

A Cataloging-in-Publication data record for this book is available from the
Library of Congress.

Typeset by the author using the TeX programming language.

ISBN 13: 978-0-8133-4079-1 (pbk)

Frontiers in Physics
David Pines, Editor

Volumes of the Series published from 1961 to 1973 are not officially numbered. The parenthetical numbers shown are designed to aid librarians and bibliographers to check the completeness of their holdings.

Titles published in this series prior to 1987 appear under either the W. A. Benjamin or the Benjamin/Cummings imprint; titles published since 1986 appear under the Westview Press imprint.

22. R. Brout — Phase Transitions, 1965

23. I. M. Khalatnikov — An Introduction to the Theory of Superfluidity, 1965

24. P. G. deGennes — Superconductivity of Metals and Alloys, 1966

25. W. A. Harrison — Pseudopotentials in the Theory of Metals, 1966

26. V. Barger / D. Cline — Phenomenological Theories of High Energy Scattering: An Experimental Evaluation, 1967

27. P. Choquàrd — The Anharmonic Crystal, 1967

28. T. Loucks — Augmented Plane Wave Method: A Guide to Performing.Electronic Structure Calculations—A Lecture Note and Reprint Volume, 1967

29. Y. Ne'eman — Algebraic Theory of Particle Physics: Hadron Dynamics In Terms of Unitary Spin Current, 1967

30. S. L. Adler / R. F. Dashen — Current Algebras and Applications to Particle Physics, 1968

31. A. B. Migdal — Nuclear Theory: The Quasiparticle Method, 1968

32. J. J. J. Kokkede — The Quark Model, 1969

33. A. B. Migdal — Approximation Methods in Quantum Mechanics, 1969

34. R. Z. Sagdeev — Nonlinear Plasma Theory, 1969

35. J. Schwinger — Quantum Kinematics and Dynamics, 1970

36. R. P. Feynman — Statistical Mechanics: A Set of Lectures, 1972

37. R. P. Feynman — Photon-Hadron Interactions, 1972

38. E. R. Caianiello — Combinatorics and Renormalization in Quantum Field Theory, 1973

39. G. B. Field / H. Arp / J. N. Bahcall — The Redshift Controversy, 1973

40. D. Horn / F. Zachariasen — Hadron Physics at Very High Energies, 1973

41. S. Ichimaru — Basic Principles of Plasma Physics: A Statistical Approach, 1973 (2nd printing, with revisions, 1980)

42. G. E. Pake / T. L. Estle — The Physical Principles of Electron Paramagnetic Resonance, 2nd Edition, completely revised, enlarged, and reset, 1973 [cf. (9)—1st edition

Volumes published from 1974 onward are being numbered as an integral part of the bibliography.

43. C. Davidson — Theory of Nonneutral Plasmas, 1974

44. S. Doniach / E. H. Sondheimer — Green's Functions for Solid State Physicists, 1974

45. P. H. Frampton — Dual Resonance Models, 1974

46. S. K. Ma — Modern Theory of Critical Phenomena, 1976

47. D. Forster — Hydrodynamic Fluctuation, Broken Symmetry, and Correlation Functions, 1975

48. A. B. Migdal — Qualitative Methods in Quantum Theory, 1977

49. S. W. Lovesey — Condensed Matter Physics: Dynamic Correlations, 1980

50. L. D. Faddeev / A. A. Slavnov — Gauge Fields: Introduction to Quantum Theory, 1980

51. P. Ramond — Field Theory: A Modern Primer, 1981 [cf. 74—2nd ed.]

52. R. A. Broglia / A. Winther — Heavy Ion Reactions: Lecture Notes Vol. I, Elastic and Inelastic Reactions, 1981

53. R. A. Broglia / A. Winther — Heavy Ion Reactions: Lecture Notes Vol. II, 1990

Frontiers in Physics
David Pines, Editor

Volumes of the Series published from 1961 to 1973 are not officially numbered. The parenthetical numbers shown are designed to aid librarians and bibliographers to check the completeness of their holdings.
Titles published in this series prior to 1987 appear under either the W. A. Benjamin or the Benjamin/Cummings imprint; titles published since 1986 appear under the Westview Press imprint.

Volumes published from 1974 onward are being numbered as an integral part of the bibliography.

Editor's Foreword

The problem of communicating in a coherent fashion recent developments in the most exciting and active fields of physics continues to be with us. The enormous growth in the number of physicists has tended to make the familiar channels of communication considerably less effective. It has become increasingly difficult for experts in a given field to keep up with the current literature; the novice can only be confused. What is needed is both a consistent account of a field and the presentation of a definite "point of view" concerning it. Formal monographs cannot meet such a need in a rapidly developing field, while the review article seems to have fallen into disfavor. Indeed, it would seem that the people who are most actively engaged in developing a given field are the people least likely to write at length about it.

Frontiers in Physics was conceived in 1961 in an effort to improve the situation in several ways. Leading physicists frequently give a series of lectures, a graduate seminar, or a graduate course in their special fields of interest. Such lectures serve to summarize the present status of a rapidly developing field and may well constitute the only coherent account available at the time. One of the principal purposes of the *Frontiers in Physics* series is to make notes on such lectures available to the wider physics community.

As *Frontiers in Physics* has evolved, a second category of book, the informal text/monograph, an intermediate step between lecture notes and formal text or monographs, has played an increasingly important role in the series. In an informal text or monograph an author has reworked his or her lecture notes to the point at which the manuscript represents a coherent summation of a newly developed field, complete with references and problems, suitable for either classroom teaching or individual study.

The study of the structure, phase behavior, and dynamics of surfaces, interfaces, and membranes has long been a central topic in colloidal science. Moreover, it has recently been realized that this sub-field of science represents a promising new direction for both statistical physicists (who are concerned with a macroscopic description) and materials scientists (who are interested in understanding and designing complex materials). Dr. Safran has been a leader in bringing a modern theoretical perspective to bear on the large-scale properties of these systems, and in bringing them to the attention of the broader physics and materials science communities. In this lecture note volume addressed to a broad interdisciplinary audience of physicists, physical chemists, chemical engineers, and materials scientists, he both provides a lucid introduction to the basic concepts of colloid science and places these in a statistical physics perspective. It is a pleasure to welcome Samuel Safran to *Frontiers in Physics*.

David Pines
Urbana, Illinois
January, 1994

Contents

Preface

The study of the structure, phase behavior, and dynamics of surfaces and interfaces is fascinating since two-dimensional interfaces "live" in and can therefore extend into the three-dimensional world, exhibiting phenomena that cannot be observed in three-dimensional, bulk materials. From a practical point of view, the engineering of tailor-made materials requires the understanding of the behavior of complex, multicomponent systems and of the internal interfaces found therein. Complex fluids and solids that are important in the development of new materials cannot be designed using trial-and-error methods due to the multiplicity of components and parameters. While these materials can sometimes be analyzed in terms of microscopic mixtures, it is often advantageous to regard them as dispersions and to focus on the properties of the internal interfaces found in these systems. This can lead to considerable simplification in the study and understanding of the system, since instead of analyzing the behavior of a complex, three-dimensional mixture, the problem is reduced to the study of the (ensemble of) two-dimensional interface(s). This simplification is applicable, of course, when the scale of the interface is much larger than that of a single molecular unit; otherwise it is best to describe the system as a bulk, three-dimensional molecular mixture.

This approach is the basis behind the theoretical presentation in this volume of *Lecture Notes on the Statistical Thermodynamics of Surfaces, Interfaces, and Membranes*. Focusing on the large-scale properties of these systems, these notes are meant to supplement the many excellent treatments of the structure and thermodynamics of surfaces, interfaces, and membranes in traditional texts on colloid science, with the concepts and theoretical tools for describing the properties of the system on scales that are large compared to a molecular size. The presentation is that of a set of lecture notes (with the attendant informality in the text, references, and figures) and the reader is referred to the more traditional texts and monographs (see the references at the end of Chapter 1) for more details concerning materials and experiments. This book is addressed to physicists, physical chemists, chemical engineers, and materials scientists who are interested in the statistical mechanics that underlies the macroscopic, thermodynamic properties of these systems. Although a previous graduate-level course on statistical mechanics is preferable, some introductory material is presented in Chapter 1 in the form of a brief review of statistical mechanics and some notions concerning the differential geometry of surfaces and hydrodynamics; almost all of the later development is self-contained once the introductory material is mastered. From experience, I have found that a one-semester, graduate-level course can cover most, but not all, of the topics discussed in the text. The text also contains worked examples (denoted by the sans serif typeface) and further problems are found at the end of each chapter.

The approach adapted here first presents the traditional treatment of these problems and then investigates how thermal fluctuations of the system affect the simple description. This philosophy is used throughout to treat the rich diversity of systems investigated in the field of colloid and interface science. The focus is generally on fluid-fluid or fluid-vapor interfaces since these systems are amenable to simple continuum descriptions, but the solid-vapor interface is discussed in the context of roughening and wetting. We begin with the properties of a single, isolated interface in Chapter 2, where we relate the surface/interfacial tension to the composition profiles of the phase separating components. Fluctuations are examined in Chapter 3, where we discuss capillary undulations of surfaces and their effect on the roughening transition. Interfaces do not exist in isolation, and the effect of three-phase contact is examined in Chapter 4 where wetting is discussed and fluctuations of the contact line are treated. Many systems of both practical and scientific interest consist of an ensemble of interfaces whose behavior is influenced by the interactions between them. Chapter 5 describes the interactions between rigid interfaces. Here, both direct as well as fluctuation-induced effects are presented in the discussions of van der Waals, electrostatic, and solute-induced interactions, sometimes known as depletion interactions. The properties and interactions of flexible interfaces (fluid membranes) are discussed in Chapter 6 with emphasis on the role of fluctuations and curvature energy. While Chapters 5 and 6 treat interactions between a small number of interfaces or surfaces, Chapters 7 and 8 discuss the thermodynamic properties of ensembles of many interfaces. Colloidal dispersions are treated in Chapter 7; here the primary interactions between the colloidal particles are via their surfaces. The problem is simplified in the sense that the structure of the surface (and of the colloidal particle) is fixed and the questions of interest concern the cooperative behavior of the ensemble. Self-assembling systems are examined in Chapter 8; here the properties of the single interface are not fixed as they are in a solidlike colloidal particle, but change as the thermodynamic conditions are varied.

While the primary focus of the book is on the systems important in colloid and interface science, a more general goal is to introduce the reader to several theoretical methods that are useful in applications of statistical mechanics to materials. Thus, in Chapter 2 a general variational method for finding the optimum "mean-field" theory applicable to a given system is presented. The extension of this method to systems that are fluctuation dominated and to the roughening transition is discussed in Chapter 3. Chapter 5 on surface interactions presents a general discussion of the stress tensor and its relation to the interactions between surfaces separated by a medium. This concept is also used in Chapter 6 to relate the bending elastic moduli of materials to their internal stress distributions, which can be calculated from microscopic models. Renormalization-group theory is not treated since it is usually restricted to an audience that is much narrower than the intended readership. Fortunately, most of the interesting physical properties discussed here can be understood in terms of theories that focus on the free energy and Gaussian fluctuations; references are made to instances where these simpler treatments break down. For similar reasons, the text mainly focuses on equilibrium properties of interfaces in relatively simple systems. Much of the rich and varied behavior of dynamically unstable interfaces is outside the scope of this book, although the dynamics of wetting and of capillary instabilities are presented. The interesting and technologically important problems related to interfaces in liquid crystalline materials are not treated; polymers at interfaces are discussed in some of the problems and in the chapter on colloid stabilization. In general, the choice of topics and the level of presentation was influenced by the notion that this volume should reach a wide, interdisciplinary audience. It is thus hoped that the depth and breadth of coverage will introduce the condensed matter physicist to colloid science and present to the physical chemist or material scientist, who may already be familiar with the underlying phenomena, a modern theoretical perspective.

I wish to thank those colleagues and students who have read and discussed these lecture notes with me. The comments of S. Alexander, D. Andelman, R. Bar Ziv, X. Chatellier, N. Dan, E. Frishman, W. Helfrich, J. F. Joanny, J. Klein, T. Lubensky, R. Menes, P. Pieruschka, P. Pincus, U. Steiner, Z. G. Wang, A. Weinstein, and T. Witten have been particularly helpful. I also am grateful to M. Cymbalista for typesetting help and to the Aspen Center for Physics for their hospitality. While these notes reached their final form through their use in graduate courses at the Weizmann Institute, my understanding of the physical properties of interfaces and membranes was developed through many stimulating interactions with my former colleagues at Exxon Research and Engineering. Finally, I wish to thank Marilyn and our children for their patience and understanding during the writing of this book and my parents and in-laws for their encouragement over the years.

Rehovot, Israel
Spring, 1994

For the paperback edition, typographical errors in both the text and equations have been corrected and some minor rewording of the text has been inserted for clarity. I am grateful for Prof. S. Komura for his careful reading of the original edition in preparation for his Japanese translation and for his suggestions.

Rehovot, Israel

Spring, 2002

CHAPTER 1

Mixtures and Interfaces

1.1 INTRODUCTION

In this chapter, we begin with some remarks on the technological and scientific importance of complex materials and interfaces and motivate the study of interface and surface properties. We then review some of the physical and mathematical methods that are used in the subsequent discussions of interface and membrane statistical thermodynamics. Many of these topics are discussed more fully in the references and throughout this chapter. We begin with a review of classical statistical mechanics[1,2,3,4], including a description of fluctuations about equilibrium and of binary mixtures. The mathematical description[5] of an interface is then presented (using only vector calculus) and the calculation of the area and curvature of an interface with an arbitrary shape is demonstrated. Finally, the chapter is concluded by a brief summary of hydrodynamics[6].

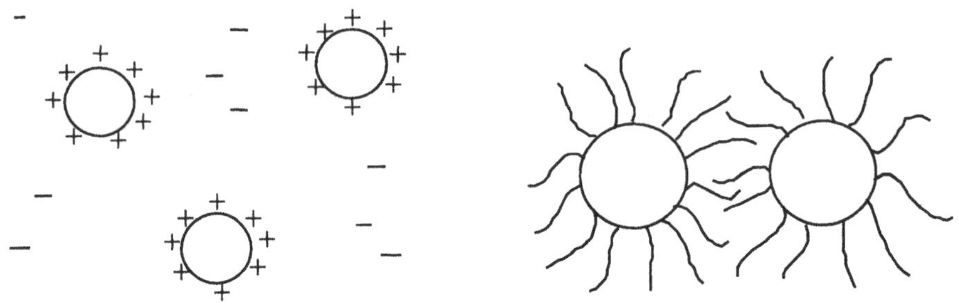

Figure 1.1 Colloidal particles in solution stabilized by electrostatic repulsion due to surface charges (left) or by steric repulsion of grafted long-chain, polymeric molecules (right).

1.2   COMPLEX MATERIALS AND INTERFACES

Complex materials refer to either composite solids or to fluid dispersions where the individual molecular species are organized into structures with long (tens of Angstroms to microns) length scales (compared with atomic or simple molecular dimensions). These materials[7,8], which include[9,10] solid colloidal dispersions (see Fig. 1.1), polymers[11,12], and self-assembled amphiphiles[13,14,15] such as micelles, vesicles, and microemulsions are important technologically and as generic models for biological systems. Milk, blood, paint, soaps, and detergents are familiar examples where the properties of dispersion, encapsulation, and cleaning are utilized. While the study of single-component systems such as simple fluids or solids focuses on the behavior of the bulk, serious consideration of the multicomponent nature of complex fluids[16] requires an understanding of the interfaces between materials. The unique properties of the dispersion are then often determined by the behavior of the interface. This often leads to considerable simplification in the study and understanding of the system, since instead of analyzing the behavior of a complex, three-dimensional mixture, the problem is reduced to the study of the two-dimensional interface. This simplification is applicable, of course, when the scale of the interface is much larger than that of a single molecular unit; otherwise it is best to describe the system as a bulk, three-dimensional molecular mixture.

The simplest example of such a system is the interface between a semi-infinite, bulk system and vacuum (or its own dilute, vapor phase); this interface is generally referred to as a **surface**. When this semi-infinite material coexists with another condensed phase, the separating surface is referred to as an **interface**. An interface can also be composed of a material that is different from the

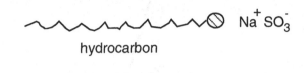

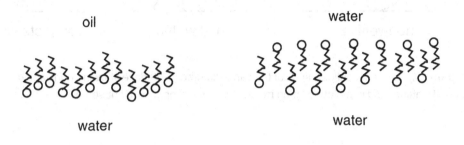

Figure 1.2 Surfactant molecule with polar head group and hydrocarbon chain. Membranes composed of an amphiphilic monolayer, separating water and oil, and an amphiphilic bilayer, separating two water regions.

two bulk phases. Examples include surface coatings and membranes composed of amphiphilic (see Fig. 1.2) (soaplike) molecules that may lie between bulk domains of water and oil. The physics of a single surface or interface is interesting and rich since the two-dimensional nature of the problem greatly magnifies the effects of thermal fluctuations on the interfacial statistical thermodynamics[17]. The understanding of single interfaces is also important in many applications such as catalysis (chemical reactions that occur on surfaces or interfaces), coatings (*e.g.,* applying a molecularly thin semiconducting or insulating coating to a metal surface), and friction and wear (where the forces between two surfaces or interfaces are of importance).

A study of the physics of single surfaces or interfaces begins with the characterization of the shape of the interface. The types of questions that one asks include: Where is the interface located? What is the free energy to create an interface? Is the interface in thermal equilibrium and thus capable of changing its structure or composition in response to changes in temperature or other system parameters, or is the interface quenched with a structure or composition that is fixed? For systems in thermal equilibrium, do thermal fluctuations cause the surface to be rough or is the interface molecularly smooth? Molecular-scale roughness is also important for characterizing the effects of coatings on surfaces. The structure of these coatings is closely related to their function and provides interesting examples of two-dimensional physics and chemistry. When these

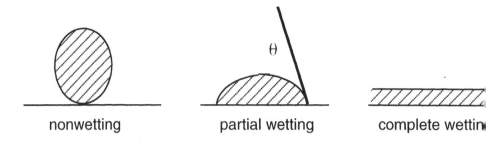

nonwetting   partial wetting   complete wetting

Figure 1.3 Different configurations of the vapor-fluid-solid interface showing nonwetting, partial wetting, and complete wetting by the fluid layer atop a solid substrate.

coatings reach quasimacroscopic (many monolayers) sizes one speaks about the wettability (see Fig. 1.3) of the surface or interface. For example, for the case of a fluid wetting a solid substrate, the important questions are: Does the fluid spread (completely wet) on the solid or does it "bead up"? What are the shapes of the fluid droplets as they spread? What is the stability of these wetting layers?

Of course, in many applications the behavior of the single, "clean" interface is insufficient to characterize the system. Applications such as enhanced oil recovery (where the behavior of water and oil in small crevices in rocks is important), inks and pigments, detergents, and cosmetics utilize colloidal dispersions (*e.g.,* liquid suspensions of either solid particles or fluid droplets with sizes from 10 Å to 10,000 Å (1 μ)). The understanding[9,10] of the structure, dynamics, and rheology (flow properties) of these systems focuses on the interactions between two (or more) interfaces since the colloidal "particles" interact mainly through their surfaces. These contacts are usually short-range compared with the interparticle spacing; in charged systems, however, the electrostatic interactions can become very long-range, resulting in anomalous properties. On the other hand, sometimes these surface-dominated interactions can be relatively weak so that the thermal energies (resulting in Brownian motion of the particle) must be considered to predict the structure, phase behavior, or rheology. Particular questions of interest include: What are the relevant interactions[18] between the surfaces? If there are strong attractive interactions, which would tend to force the surfaces to adhere and thus destabilize a colloidal suspension, can adsorbed species like grafted polymers help change the interactions from attractive to repulsive and thus stabilize the suspension? What are the interactions between such surfaces

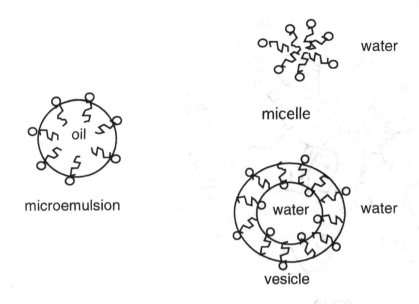

water

micelle

oil

microemulsion

water

water

vesicle

Figure 1.4 Self-assembly of amphiphiles into micelles, vesicles, and microemulsions.

— both when at rest and when in motion? What is the phase behavior of colloidal suspensions as a function of the surface treatment?

Finally, an even richer set of phenomena and applications is related to the behavior of *self-assembling colloids* as shown in Fig. 1.4. In such systems, generally composed of[13,14,15] (soaplike) molecules in solution, the colloidal "particles" are not rigid; their size and shape are influenced by forces comparable to those that exist between them and can be "tuned" by varying the temperature, solvent, or molecular chemistry of the surfactant. The surfactant molecules are characterized by a polar section that tends to reside in a high dielectric constant medium, like water, and a hydrocarbon part that is relatively insoluble in water. These molecules thus tend to aggregate at air-water or oil-water interfaces with a definite directionality: the polar part in the water and the hydrocarbon section in the oil. The structures formed by these aggregates show a variety of forms and properties ranging from ordered arrays of micelles (*e.g.,* spherical aggregates of surfactant with their polar parts on the sphere surface and solubilized in the water) to disordered, bicontinuous microemulsions (see Fig. 1.5) (dispersions of water and oil with surfactant monolayers at the interfaces between the water and oil domains).

In addition to short-chain (≤ 20 hydrocarbon groups) surfactants, long-chain polymers can sometimes show surfactant-like behavior. Examples[19] of these are

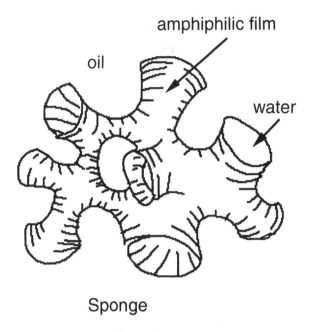

Figure 1.5 Representation of a spongelike structure of a bicontinuous microemulsion that consists of a surfactant monolayer separating oil and water domains.

polymers with polar groups attached at one end, and block copolymers, where two or more *incompatible* polymers are chemically joined (see Fig. 1.6). Both surfactants and block copolymers are important model systems for biological structures such as cell walls, membranes, and lining materials of the lungs. In addition, the microstructures formed by self-assembling colloids are useful in the design of smart microcapsules for the pharmaceutical industry[20]. The important conceptual questions related to these systems include the shape and size distribution of the self-assembled aggregates, their phase behavior as a function of temperature and solvent quality, and their dynamics and rheology.

These chapters present a theoretical discussion of the structural and thermodynamic properties of single surfaces and interfaces, interacting rigid interfaces important in colloid science, and self-assembling interfaces composed of surfactants and polymers. In these studies of the physical properties of complex fluids, the emphasis is on the behavior of systems at "mesoscopic" length scales of tens to thousands of Angstroms and the role of thermal fluctuations, correlations, randomness, and long-range interactions is addressed. This approach is emphasized in our discussion of interfaces and colloids, which studies the average (mean-field) configurations and then investigates the effects of fluctuations about these

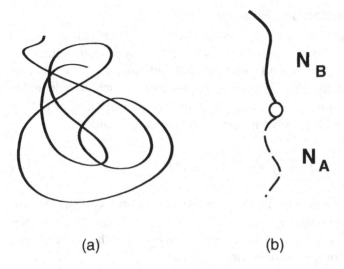

(a)   (b)

Figure 1.6 (a) Long-chain polymeric molecule in a good solvent; the configuration of the chain resembles that of a random walk. (b) Block copolymer molecule; where N_A and N_B represent the molecular weights of the two blocks that are joined together with a chemical bond, represented by the open circle.

average structures. For simplicity and generality we focus on fluid-fluid, fluid-solid, and fluid-vapor interfaces. Although this book presents a theoretical view of the structure and phase behavior of interfaces and membranes, the questions raised are very relevant to experimental studies that use neutron, light, and X-ray scattering, electron microscopy, or real-space nuclear-reaction analysis, to measure the structure, phase behavior and dynamics of these materials.

1.3   REVIEW OF CLASSICAL STATISTICAL MECHANICS

This section summarizes the classical, equilibrium, statistical mechanics of many-particle systems, where the particles are described by their positions, $\vec{q}$, and momenta, $\vec{p}$. The section begins with a review of the definition of entropy and a derivation of the Boltzmann distribution and discusses the effects of fluctuations about the most probable state of a system. Some worked examples are presented to illustrate the thermodynamics of the nearly ideal gas and the Gaussian probability distribution for fluctuations.

Probability Distribution and Entropy

The key ingredient in a statistical description of classical, many-particle systems where each particle labeled by i is described by its momentum, $\vec{p}_i$, and position, $\vec{q}_i$, is the **probability distribution function**, $P(\{\vec{p}_i, \vec{q}_i\})$. This distribution is a function of the momenta and positions of each particle and is taken to be normalized to unity — *i.e.*, the sum of the probabilities of finding any given particle with all possible momenta and positions is unity:

$$\int d\vec{p}_i \int d\vec{q}_i \, P(\{\vec{p}_i, \vec{q}_i\}) = 1 \qquad (1.1)$$

where the integrals extend over the entire phase space of the system. The importance of $P(\{\vec{p}_i, \vec{q}_i\})$ is that once it is known, one can calculate the ensemble-averaged value, denoted by $\langle ... \rangle$, of any observable property of the system. Thus, if A is some function of the momenta and positions,

$$\langle A \rangle = \prod_i \int d\vec{p}_i \int d\vec{q}_i \, A(\vec{p}_i, \vec{q}_i) \, P(\{\vec{p}_i, \vec{q}_i\}) \qquad (1.2)$$

The entropy, S, is *defined* by the negative of the average value of the logarithm of $P(\{\vec{p}_i, \vec{q}_i\})$ and is a measure of the number of states that are available to the system:

$$S = -\prod_i \int P(\{\vec{p}_i, \vec{q}_i\}) \log[P(\{\vec{p}_i, \vec{q}_i\})] \, d\vec{p}_i \, d\vec{q}_i \qquad (1.3)$$

In thermal equilibrium, the most probable distribution is the one that maximizes the entropy, subject to any constraints that may act on the system. These results are easily generalized to systems where the degrees of freedom are not necessarily those of position and momentum — *e.g.*, two-state systems. For simplicity, in the following discussion, we consider position and momentum. Also for simplicity, we drop the index i and consider the vectors $\vec{p}$ and $\vec{q}$ to denote the positions and momenta of all of the particles: *e.g.*, $\vec{p} = (p_{x,1}, p_{y,1}, p_{z,1}, p_{x,2}, p_{y,2}, p_{z,2}...)$, where 1,2... are the labels of the individual particles.

Gibbs Distribution

A system in equilibrium with a reservoir has its temperature fixed by interactions with the reservoir. If one considers both the system and the reservoir, the total energy is conserved. Thus the equilibrium probability distribution is one that maximizes the entropy of the system with respect to the constraint that the total energy of the system plus reservoir is conserved. To do this, one considers the Helmholtz free energy, F, which is defined by $F = U - TS$, where T is a constant, defined as the temperature. We shall usually use units where the Boltzmann constant, k_B, is set equal to unity. The average internal energy is denoted by U; it is given by the ensemble average of the Hamiltonian: $U = \langle \mathcal{H} \rangle$, where $\mathcal{H}(\vec{p}, \vec{q})$ is the Hamiltonian of the system (energy as a function of the coordinates and momenta of each particle),

$$U = \langle \mathcal{H} \rangle = \frac{\int d\vec{p} \, d\vec{q} \, \mathcal{H} P(\vec{p}, \vec{q})}{\int d\vec{p} \, d\vec{q} P(\vec{p}, \vec{q})} \tag{1.4}$$

and S is the entropy of the system given by Eq. (1.3). To determine the most probable distribution for the system we note that both S and U and hence F are functionals of the probability distribution function. One thus determines this distribution function by maximizing the entropy of the system and reservoir with respect to the *function* $P(\vec{p}, \vec{q})$, subject to the constraints that (i) the total energy of the system plus reservoir is fixed and (ii) the normalization of the probability distribution (Eq. (1.1)). We imagine that the reservoir, which is much larger than the system, is not affected by the system. In addition, we also neglect the fluctuations in energy of the system, and replace the constraint of fixed total energy by the constraint that the average energy, U, of the system is fixed. Writing $-F/T = S - (1/T)U$, one can see that maximizing the entropy is equivalent to *minimizing* the free energy, F (because of the negative sign relating F and S) where $1/T$ is a Lagrange multiplier that accounts for the conservation of the energy. Using Eqs. (1.3,1.4), one can show that $\delta F/\delta P = 0$ implies that the probability distribution satisfies the Boltzmann relation:

$$P(\vec{p}, \vec{q}) \sim e^{-\mathcal{H}(\vec{p}, \vec{q})/T} \tag{1.5}$$

where $\mathcal{H}$ is the Hamiltonian. The proportionality constant is obtained by enforcing the normalization condition, Eq. (1.1).

Partition Function

For a system with a single particle in three spatial dimensions, the **partition function**, Z, is defined by

$$Z = \int \frac{d\vec{p} \, d\vec{q}}{(2\pi\hbar)^3} e^{-\mathcal{H}/T} \tag{1.6}$$

where the factor of $(2\pi\hbar)^3$ is a normalization of Z that takes into account[1] the quantum mechanical uncertainty principle; the resulting Z is thus dimensionless.

Since all physical observables are related to averages with the probability distribution function, this factor cancels when the normalized distribution function is used; it therefore does not enter in any of the expressions for the observables in the thermodynamic limit of very large systems. We shall take this factor into account by defining the phase space:

$$dA = \frac{d\vec{p}\, d\vec{q}}{(2\pi\hbar)^3} \tag{1.7}$$

For a system with N indistinguishable particles[2] we write the partition function explicitly as

$$Z = \frac{1}{N!}\int dA\, e^{-\mathcal{H}[\vec{p},\vec{q}]/T} \tag{1.8}$$

where the factor of $1/N!$ accounts for the indistinguishability of the particles[2] and where the integrals over $\vec{p}$ and $\vec{q}$ include all the particle coordinates:

$$dA = \prod_{i=1}^{N} \frac{d\vec{p}_i\, d\vec{q}_i}{(2\pi\hbar)^3} \tag{1.9}$$

From these definitions and from Eqs. (1.3,1.4), one can show[2] that the free energy,

$$F = -T\log Z \tag{1.10}$$

Similarly, one can show that the entropy is

$$S = -\frac{\partial F}{\partial T} \tag{1.11}$$

and the internal energy is

$$\mathcal{U} = \frac{\partial(F/T)}{\partial(1/T)} \tag{1.12}$$

These expressions assume that the total number of particles in the system is fixed. For a system with a thermodynamically large number of particles ($N \to \infty$, so one can neglect fluctuations in the particle number), one can equally well minimize the free energy with the constraint that only the *average* number of particles is fixed. One therefore considers the grand potential, G, given by

$$G = F - \mu\bar{N} \tag{1.13}$$

where μ is the chemical potential — a Lagrange multiplier that fixes the average number of particles $\bar{N}$ given by

$$\bar{N} = \sum_{M=1}^{\infty} \int M \, P_M(\vec{p}, \vec{q}) \, d\Lambda \qquad (1.14)$$

where the sum is over over all possible numbers of particles and P_M denotes the probability distribution function for a system with M particles:

$$P_M(p, q) \sim \frac{1}{M!} \, e^{-\mathcal{H}_M(\vec{p}, \vec{q})/T + \mu M/T} \qquad (1.15)$$

where the Lagrange multiplier enforces the requirement that the average number of particles is equal to $\bar{N}$. The modified partition function, $Z_{\bar{N}}$, is

$$Z_{\bar{N}} = \sum_{M=1}^{\infty} \frac{1}{M!} \int e^{-\mathcal{H}_M(\vec{p}, \vec{q})/T + \mu M/T} d\Lambda \qquad (1.16)$$

Averages are calculated by integrating over $(\vec{p}, \vec{q})$ and summing over M. The chemical potential is determined by requiring that Eq. (1.14) be satisfied and can also be written as

$$\bar{N} = \frac{T}{Z_{\bar{N}}} \frac{\partial Z_{\bar{N}}}{\partial \mu} = -\frac{\partial G}{\partial \mu} \qquad (1.17)$$

It is important to note that for classical systems, the momenta, $\vec{p}$, and the positional degrees of freedom, $\vec{q}$, are decoupled. One can therefore integrate over the momenta to obtain an effective partition function which depends only on $\vec{q}$. The *equilibrium* behavior of the system is then determined by considering only the positional degrees of freedom. This can be seen in the following example, where both the momentum and position are considered, but where the effect of the integration over the momenta is to add a constant term to the free energy. In most of this book (but see Chapters 3 and 4), we therefore omit consideration of the kinetic energy of the system.

Example: Nearly Ideal Gas

Consider a nearly ideal gas that is dilute, so the free energy can be expanded in a power series in the density, n. Derive expressions for the free energy up to second order in n for particles that interact with a potential $U(\vec{r} - \vec{r}\,')$.

To derive this expansion consider the grand potential:

$$G = F - \mu \bar{N} \tag{1.18}$$

where μ is the chemical potential that fixes the correct number of particles, $\bar{N}$, $(\partial G/\partial \mu = -\bar{N})$. For a system with fixed μ, one finds G by integrating over both the state variables (positions $\vec{q}$, momenta $\vec{p}$) and summing over the number of particles.

$$G = -T \log \left\{ \sum_{M=0}^{\infty} \frac{1}{M!} e^{\mu M/T} \int e^{-\mathcal{H}_M(p,q)/T} \, d\Lambda \right\} \tag{1.19}$$

where $\mathcal{H}_M$ is the energy Hamiltonian for a system with M particles in the system and the energies, E_M, are

$$E_0 = 0 \tag{1.20a}$$

$$E_1 = p^2/2m \tag{1.20b}$$

$$E_2 = p_1^2/2m + p_2^2/2m + U(\vec{r}_1 - \vec{r}_2) \tag{1.20c}$$

since the interaction potential is only relevant when there are two or more particles in the system. The phase-space factor $d\Lambda$ is defined in Eq. (1.9).

To proceed with the expansion we assume that the gas is dilute and consider a volume, V, which is large compared to atomic distances, but is small compared to the total volume of the system; we thus consider a region where the average number of particles in the volume V is small. Since the gas is dilute and the particles are indistinguishable, we can construct the total free energy by simply adding together the contributions of many such regions; this is because the free energy is extensive. Using this notation and keeping terms up to order $M = 2$, one obtains an expansion up to quadratic order in $\xi = e^{\mu/T}/v_0$ where

$$v_0 = \left(\frac{2\pi\hbar^2}{mT} \right)^{\frac{3}{2}} \tag{1.21}$$

is a characteristic volume that comes from the integration over the momenta.
Then:

$$G = -T \log \left[1 + \xi V + \frac{\xi^2}{2} \int e^{-U(r-r')/T} d\vec{r} d\vec{r'} \right] \qquad (1.22)$$

For the ideal gas, ξ is proportional to the density, so we attempt a low density
expansion of the free energy by expanding the logarithm to second order in
ξ. Solving for μ from the constraint yields an expansion for the free energy
up to second order in the density:

$$G = -TV \left[\xi - \tfrac{1}{2}\xi^2 a... \right] \qquad (1.23)$$

where

$$a = \int d\vec{r} \left(1 - e^{-U(\vec{r})/T} \right) \qquad (1.24)$$

We find

$$n \approx \xi - a\xi^2; \quad \xi \approx n(1 + an) \qquad (1.25)$$

Thus, keeping terms only to order n^2 and substituting for ξ we find the free
energy, $F = G + \mu\bar{N}$:

$$F = T\bar{N} \left[(\log[nv_0] - 1) + \tfrac{1}{2}an \right] \qquad (1.26)$$

$$F = TV \left[n(\log[nv_0] - 1) + \tfrac{1}{2}an^2 \right] \qquad (1.27)$$

The first term in the free energy is from the entropy of the ideal gas, while
the second term takes into account both the interactions and the entropy of
the configurations.

The constant, a, is known as the **virial coefficient**. It is the coefficient
of the lowest order term in an expansion of the free energy as a function
of the density, that takes into account the interparticle interactions. When
$a > 0$ the interactions are predominantly repulsive and when $a < 0$, they
are predominantly attractive. We note that this "virial expansion" of the free
energy is correct at low densities, since it is an expansion to quadratic order
in n. ■

Density Functional Theory

The derivation of the free energy of a homogeneous, nearly ideal gas motivates the following heuristic treatment of interacting particles with spatial inhomogeneity, such as that found near a surface. For "small" or more accurately, slowly varying, inhomogeneities in the density, one attempts to write the free energy as a functional of the spatially dependent density, $n(\vec{r})$, of the system and to determine the density profile in equilibrium by functionally minimizing this free energy with respect to $n(\vec{r})$. It is assumed that the form of the free energy as a function of $n(\vec{r})$ is related to the free energy as a function of the density for the homogeneous case; the boundary conditions will then dictate how the non-homogeneous aspects enter and the nonlocality of the potential will dictate how these effects propagate. Thus, for a system that is nearly ideal and is described by a local density $n(\vec{r})$, one writes:

$$F = T \int d\vec{r}\, n(\vec{r}) \left[\log[n(\vec{r})v_0] - 1\right] + \tfrac{1}{2}T \int d\vec{r}\, d\vec{r}'\, n(\vec{r})\, n(\vec{r}')\, U_e(\vec{r} - \vec{r}')$$

(1.28a)

where the effective potential, U_e is

$$U_e(\vec{r} - \vec{r}') = \left[1 - e^{-U(\vec{r} - \vec{r}')/T}\right]$$

(1.28b)

which reduces to the nonideal gas free energy of Eq. (1.27) in the homogeneous limit. If $|U|/T \ll 1$, the entropy and energy terms are separable and

$$F = T \int d\vec{r}\, n(\vec{r}) \left[\log[(n(\vec{r})v_0] - 1\right] + \tfrac{1}{2} \int d\vec{r}\, d\vec{r}'\, n(\vec{r})\, n(\vec{r}')\, U(\vec{r} - \vec{r}') \quad (1.29)$$

To determine the average density, thus neglecting fluctuations, we minimize the free energy with respect to the density $n(\vec{r})$ with the constraint that the total density is constant. Noting that the interaction term has a "quadratic" term in $n(\vec{r})$, we get a nonlocal equation for the density:

$$T \log[n(\vec{r})v_0] + \int d\vec{r}'\, n(\vec{r}')\, U(\vec{r} - \vec{r}') = \mu$$

(1.30)

where μ is a chemical potential that enforces constant total density.

If the interaction is short-range, we can expand $n(\vec{r}')$ in a series around $n(\vec{r})$. Noting that U is usually symmetric in $\vec{r}$ and $\vec{r}'$ so that

$$\int d\vec{r}'(\vec{r} - \vec{r}')\, U(\vec{r} - \vec{r}') = 0$$

(1.31)

we have to keep up to second-order terms in the expansion in $n(\vec{r})$. The minimization equation then reads:

$$\bar{U} \, n(\vec{r}) + \log[n(\vec{r})v_0] - \xi_0^2 \nabla^2 \left[n(\vec{r})v_0\right] = \mu' \tag{1.32}$$

where μ' is a constant, independent of space and

$$\xi_0^2 = -\frac{1}{2Tv_0} \int d\vec{r} \; \vec{r}^2 \, U(\vec{r}) \tag{1.33}$$

$\bar{U} = \int d\vec{r} \, U(\vec{r})/T$ and does not necessarily vanish. For an attractive potential, $\xi_0 > 0$. The linear moments of the interaction potential vanish by symmetry. The length scale ξ_0 is known as the bare correlation length of the potential. For example, for a potential with an exponential decay length λ, (*i.e.*, $U \sim \exp(-r/\lambda)$), the correlation length ξ_0 is proportional to λ.

However, the identification of U with the bare interaction potential is correct only in the very low density limit. At higher densities, the potential U that enters the free energy must account for indirect interactions between pairs of particles separated by even relatively large distances. The relevant quantity is an effective potential related to the pair distribution function of the system. Applications of these ideas to liquids and to liquid surfaces, using detailed, microscopic models are discussed in a unified manner in Ref. 21. In the treatment presented here, we shall focus on a more phenomenological approach.

Thermal Averages and Fluctuations

We now consider the calculation of the thermal averages and fluctuations of various functions of the position and momenta. As in Eq. (1.2), we define the average value of an operator $A(\vec{p}, \vec{q})$ to be

$$\langle A \rangle = \frac{1}{Z} \int d\Lambda \; e^{-\mathcal{H}(\vec{p},\vec{q})/T} \, A(\vec{p}, \vec{q}) \tag{1.34}$$

where $A(\vec{p}, \vec{q})$ is any function of the position and momenta and $d\Lambda$ is the phase-space integral given in Eq. (1.9). Simple examples of the the operator A are $\vec{q}, \vec{p}$ or the mean-square positions or momenta. In many-particle systems, described by a set of positions ($\vec{q}_i, i = 1..N$) and momenta ($\vec{p}_i, i = 1..N$), one may also be interested in the correlations of the positions of two different particles, *e.g.*, $\langle \vec{q}_i \cdot \vec{q}_j \rangle$.

To calculate these types of expectation values, we consider the augmented distribution function,

$$P(\vec{p}, \vec{q}, \lambda) \sim e^{-\mathcal{H}(\vec{p},\vec{q})/T - \lambda A(\vec{p},\vec{q})/T} \tag{1.35}$$

This gives rise to a free energy that is a function of λ by Eq. (1.6). The average value of A is given by

$$\langle A \rangle = \left(\frac{\partial F}{\partial \lambda} \right)_{\lambda=0} \tag{1.36}$$

The mean-square deviation of A from its average value, $\langle A \rangle$, is given by

$$\chi = \langle (A - \langle A \rangle)^2 \rangle = \langle A^2 \rangle - \langle A \rangle^2 = -T \left(\frac{\partial^2 F}{\partial \lambda^2} \right)_{\lambda=0} \tag{1.37}$$

evaluated for $\lambda = 0$.

The probability distribution for the operator $A(\vec{p}, \vec{q})$ to have a specific value, A_0 is obtained by integrating out all the other degrees of freedom:

$$P[A(\vec{p}, \vec{q}) = A_0] = \frac{1}{Z} \int \delta\left(A(\vec{p}, \vec{q}) - A_0 \right) e^{-\mathcal{H}/T} d\Lambda \tag{1.38}$$

Here, $\delta(x)$ is the Dirac delta function. One sees that this expression for the probability distribution is correct if one integrates Eq. (1.38) over all values of A_0; the result is unity. In practical applications of this formula, it is useful to evaluate the delta function using the relation:

$$\delta(x) = \frac{1}{2\pi} \int_{-\infty}^{\infty} d\omega \; e^{i\omega x} \tag{1.39}$$

allowing this exponential to be combined with the exponential arising from the Boltzmann factor.

Harmonic Oscillator

As a simple example, consider a harmonic oscillator in equilibrium. The Hamiltonian is given by $\mathcal{H} = p^2/2m + \frac{1}{2}K(q - q_0)^2$. This gives rise to a Gaussian distribution function. Using Eqs. (1.34,1.37) we find that the average position and mean-square fluctuation are given by $\langle q \rangle = q_0$ and $\langle (q - q_0)^2 \rangle = T/K$

respectively. From Eq. (1.38), we see that the probability to find the system with a position that is *not* the average position, q_0, *i.e.*, $P[q = q_0 + \delta q]$ is given by

$$P[q = q_0 + \delta q] \sim e^{-\delta q^2 / 2\sigma^2} \qquad (1.40)$$

where $\sigma^2 = T/K$.

Example: Gaussian Distribution

A generalization of the harmonic oscillator to a system with many spatial degrees of freedom is called the Gaussian model. In its simplest form, a variable, $-\infty < h(\vec{r}) < \infty$ describes the degree of freedom — for example, the position of an interface that can wander in space — and the Hamiltonian is written:

$$\mathcal{H} = \frac{1}{2} \int d\vec{r} \int d\vec{r}' h(\vec{r}) G(\vec{r} - \vec{r}') h(\vec{r}') \qquad (1.41)$$

What is the mean-square fluctuation of the variable $h(\vec{r})$?

Equation (1.37) cannot be used directly since the degrees of freedom at positions $\vec{r}$ and $\vec{r}'$ are coupled and the integrals cannot be done. However, for translationally invariant systems, the coupling only depends on the *difference* $\vec{r} - \vec{r}'$ and the Hamiltonian can be diagonalized in Fourier space. One thus considers the transforms:

$$h(\vec{r}) = \frac{1}{\sqrt{L^d}} \sum_{\vec{q}} h(\vec{q}) e^{-i\vec{q}\cdot\vec{r}} \qquad (1.42a)$$

$$h(\vec{q}) = \frac{1}{\sqrt{L^d}} \int d\vec{r} \, h(\vec{r}) e^{i\vec{q}\cdot\vec{r}} \qquad (1.42b)$$

We *define* the transform of the coupling, $G(\vec{r} - \vec{r}')$ without the factor of $\sqrt{1/L^d}$; for a translationally invariant system the coupling is a function of a single distance variable, and we write

$$G(\vec{q}) = \int d\vec{r} \, G(\vec{r}) e^{i\vec{q}\cdot\vec{r}} \qquad (1.43)$$

For systems with inversion symmetry, $G(\vec{q})$ is an even function of $\vec{q}$. The wavevector $\vec{q}$ has a discrete number of values for a system of spatial dimension d and of size L, discretized into N^d cells; each cell has a size a given by $a = L/N$, where d is the dimensionality of space relevant to the variable

h (e.g., for a surface coordinate, $d = 2$). If one uses periodic boundary conditions one finds that this set of values determines each component of the wavevector, (e.g., for a surface, $\vec{q} = (q_x, q_y)$) to obey

$$\{q_m a\} = \left\{\frac{2\pi m}{N}\right\}_{m=-N/2,...,N/2} \tag{1.44}$$

In the continuum limit $a \to 0$, the sums are converted to integrals over $\vec{q}$ by

$$\sum_{\vec{q}} = \left(\frac{L}{2\pi}\right)^d \int d\vec{q} \tag{1.45}$$

where d is the appropriate space dimensionality for the $\vec{q}$. Using these expressions in the Hamiltonian we find that

$$\mathcal{H} = \tfrac{1}{2} \sum_{\vec{q}} h(\vec{q})G(\vec{q})h(-\vec{q}) \tag{1.46}$$

The Hamiltonian is diagonal in Fourier space, except for the coupling between $\vec{q}$ and $-\vec{q}$. (This coupling is shown below to result in the fact that $\mathcal{H}$ depends only on the amplitude $|h(\vec{q})|^2$.) Thus by inspection, $\langle h(\vec{q}) \rangle = 0$ and the mean-square fluctuation of a mode with wavevector $\vec{k}$, is given by

$$\left\langle |h(\vec{k})|^2 \right\rangle = \prod_{\vec{q}} \int dh(\vec{q})\, h(\vec{k})\, h(-\vec{k})\, P[h(\vec{q})] \tag{1.47}$$

with

$$P[h(\vec{q})] = \frac{e^{-\mathcal{H}/T}}{\prod_{\vec{q}} \int dh(\vec{q})\, e^{-\mathcal{H}/T}} \tag{1.48}$$

with $\mathcal{H}$ given by Eq. (1.46). The product in Eq. (1.47) is over all the wavevectors $\{\vec{q}\}$, including $\vec{q} = \vec{k}$.

The coupling between $\vec{q}$ and $-\vec{q}$ is dealt with by noting that while $h(\vec{r})$ is a *real* quantity, as appropriate to a position variable, $h(\vec{q})$ is complex with real and imaginary parts denoted by $Re[h(\vec{q})]$ and $Im[h(\vec{q})]$ respectively. Using the definition of the Fourier transform and the fact that $h(\vec{r})$ is real, implies that

$$Re[h(\vec{q})] = Re[h(-\vec{q})] \tag{1.49a}$$

and

$$Im[h(\vec{q}\,)] = -Im[h(-\vec{q}\,)] \qquad (1.49b)$$

Thus the transforms at $\vec{q}$ and $-\vec{q}$ are *not* independent. However, one can rewrite the integral in Eqs. (1.47,1.48) as

$$\prod_{\vec{q}} \int dh(\vec{q}\,) = \int \prod_{\vec{q}>0} dRe[h(\vec{q}\,)] \, dIm[h(\vec{q}\,)] \qquad (1.50)$$

(where $\vec{q} > 0$ signifies the upper half-space of the d dimensional space of the wavevector) and use the fact that $G(\vec{q}\,) = G(-\vec{q}\,)$ (which is the case when $G(\vec{r} - \vec{r}\,') = G(\vec{r}\,' - \vec{r}\,)$) to write:

$$\sum_{\vec{q}} h(\vec{q}\,)G(\vec{q}\,)h(-\vec{q}\,) = \sum_{\vec{q}} G(\vec{q}\,)\{Re[h(\vec{q}\,)]^2 + Im[h(\vec{q}\,)]^2\} \qquad (1.51)$$

to perform the integrals over the real and imaginary parts separately for $\vec{q} > 0$. One can then write

$$\langle |h(\vec{q}\,)|^2 \rangle = -\frac{2T}{G(\vec{q}\,)} \left[\frac{\partial}{\partial \alpha} \log \int dh(\vec{q}\,) \, e^{-\frac{1}{2T}\alpha G(\vec{q}\,)|h(\vec{q}\,)|^2} \right]_{\alpha=1} \qquad (1.52a)$$

$$\langle |h(\vec{q}\,)|^2 \rangle = \frac{T}{G(\vec{q}\,)} \qquad (1.52b)$$

This is just a restatement of the equipartition theorem[1,2] for a system with many degrees of freedom. Similarly, one can easily show that there is no coupling between modes with different values of $|\vec{q}|$ and that in general:

$$\langle h(\vec{q}\,)h(\vec{q}\,') \rangle = \frac{T}{G(\vec{q}\,)}\delta_{\vec{q},-\vec{q}'} \qquad (1.53)$$

where for discrete $\vec{q}$ modes we use a Kronecker delta function. To evaluate the mean-square fluctuation in real space, one uses the definitions of the Fourier transforms and Eq. (1.53) (which changes the double sum on the two values of the wavevector to a single sum) to show that

$$\langle h^2(\vec{r}\,) \rangle = \frac{1}{L^d} \sum_{\vec{q}} \langle |h(\vec{q}\,)|^2 \rangle = \left(\frac{1}{2\pi}\right)^d \int d\vec{q}\, \frac{T}{G(\vec{q}\,)} \qquad (1.54)$$

■

Scattering Structure Factor

In a many-particle system, fluctuations of the density lead to scattering of radiation: light, neutrons, X-rays. Density fluctuations lead to polarization fluctuations, which lead to scattering[22,23]. One can define the density operator,

$$n(\vec{r}) = \sum_i \delta(\vec{r} - \vec{r}_i) \qquad (1.55)$$

where the sum is over the position of all the particles labeled by i at positions, $\vec{r}_i$. The scattering intensity, $I(\vec{q})$, is a function of the scattered wavevector, $\vec{q} = \vec{q}_{out} - \vec{q}_{in}$:

$$I(\vec{q}) = \tilde{I}_0 \int d\vec{r} d\vec{r}\,' \left\langle n(\vec{r}) n(\vec{r}\,') \right\rangle \; e^{-i\vec{q}\cdot(\vec{r}-\vec{r}\,')} \qquad (1.56)$$

where $\tilde{I}_0$ is a constant (independent of $\vec{q}$) that depends on the scattering mechanism. Another useful formula for the scattering intensity is

$$I(\vec{q}) = I_0 \left\langle n(\vec{q}) n(-\vec{q}) \right\rangle \qquad (1.57)$$

where I_0 is proportional to $\tilde{I}_0$ and $n(\vec{q})$ is the Fourier transform of the density:

$$n(\vec{q}) = \frac{1}{\sqrt{L^d}} \int d\vec{r} \, n(\vec{r}) \, e^{-i\vec{q}\cdot\vec{r}} \qquad (1.58)$$

The scattering from a system of noninteracting spheres and from a system of polymers (long, chainlike molecules) is discussed in the problems at the end of this chapter.

1.4   PHASE SEPARATION IN BINARY MIXTURES

Phase Separation and Interfaces

The previous section reviewed the elements of statistical mechanics that are important in thinking about the structures, fluctuations, and phase behavior of surfaces, interfaces, and membranes. In this section, we consider an important application of these ideas to the problem of phase separation in binary mixtures. This problem is analogous to other types of phase transitions, such as those found in Ising magnets[2]. It is important to understand the specific problem of phase separation because it is this phenomenon that results in the equilibrium between two coexisting states, which naturally gives rise to the existence of interfaces.

Mixtures

To describe the interface between two components (or a surface between a substance and its vapor or a substance and a vacuum) one first begins with the Hamiltonian and free energy of the bulk system. For generality, we consider a binary mixture of two components; they could be atoms and vacancies appropriate to describing a liquid or solid in equilibrium with its vapor or two different molecular species, as in a binary mixture. Since we are interested in thermodynamic properties and in structural properties on length scales much larger than molecular sizes, we can take into account the excluded volume interactions between the hard cores of the molecules by putting the two species on a lattice, with sites labeled by $i = 1...N$. Again, for simplicity we consider the case where the two molecular sizes are the same so there is a single lattice size in the problem. In this "lattice-gas" model we use a variable $s_i = 1$ to denote that species "B" (the solute) occupies lattice site i; $s_i = 0$ if the site is occupied by an "A" molecule (the solvent). For simplicity, we consider the case where there are only two-body interactions between the two species. We can write J_{ij}^{AA} as the interaction energy between two "A" molecules separated by a distance $|\vec{R}_i - \vec{R}_j|$. Similarly, the interactions between two "B" molecules are denoted as J_{ij}^{BB} and those between an AB pair are written J_{ij}^{AB}. The microscopic Hamiltonian that is the sum of all of these energies is thus written:

$$\mathcal{H} = -\tfrac{1}{2} \sum_{ij} K_{ij} \qquad (1.59a)$$

where

$$K_{ij} = \left[J_{ij}^{AA}(1 - s_i)(1 - s_j) + J_{ij}^{BB} s_i s_j + J_{ij}^{AB} \left[(s_i(1 - s_j) + s_j(1 - s_i)) \right] \right] \qquad (1.59b)$$

The negative sign indicates that these interactions are attractive; the J_{ij} are positive quantities that are a measure of the magnitude of the attractive interactions. The first term is nonzero only when s_i and s_j are both zero, indicating the presence of two "A" molecules on sites separated by a distance $|\vec{R}_i - \vec{R}_j|$; similar considerations apply to the second and third terms.

One can multiply all the occupation variables $\{s_i\}$ and collect terms in powers of $\{s_i\}$. The result yields a constant term that just redefines the zero of energy for the system, a linear term that multiplies $\sum_i s_i$, the average volume fraction of solute molecules (which is either fixed or determined by a chemical potential), and a quadratic term. Adding and subtracting terms linear in $\sum s_i$, the net interaction in the system can be written

$$\frac{1}{2} \sum_{ij} \left[J_{ij}^{AA} + J_{ij}^{BB} - 2J_{ij}^{AB} \right] s_i(1 - s_j) \tag{1.60}$$

Thus, if the magnitude of the attractive interactions between AA and BB pairs are larger than those between the AB pairs, the system will tend to phase separate and maximize the number of AA pairs and BB pairs in the two coexisting phases. One therefore defines

$$J_{ij} = \left[J_{ij}^{AA} + J_{ij}^{BB} - 2J_{ij}^{AB} \right] \tag{1.61}$$

and writes the interaction Hamiltonian as

$$\mathcal{H} = \frac{1}{2} \sum_{ij} J_{ij} s_i(1 - s_j) \tag{1.62}$$

To calculate the partition function, one includes a chemical potential to conserve the number of "B" particles (all of the other terms linear in the $\{s_i\}$ are included here). The constrained partition function, Z_μ, is

$$Z_\mu = \prod_k \sum_{\{s_k=0,1\}} e^{-\mathcal{H}'/T} \tag{1.63}$$

where

$$\mathcal{H}' = \mathcal{H} - \sum_i \mu s_i \tag{1.64}$$

and where the product in Eq. (1.63) applies to the sum over each of the values of $s_k = 0, 1$. All the sites are coupled to each other by the interaction and the partition function is not analytically tractable in general, although there are analytic solutions[1,2] in one and two dimensions for simple forms of the interaction matrix, J_{ij}. In the following section we discuss the system in terms of the mean-field or random-mixing approximation. This approximation can be shown to be exact in the limit of infinitely long-range interactions. A more rigorous derivation of the mean-field approximation from a variational principle is presented in chapter 2.

Mean-Field Theory

If all of the J_{ij} are positive, the system will tend to phase separate and one can consider the energy of a lattice site surrounded by its average interaction with its surroundings. Mathematically, this allows a decoupling of the interaction term involving only single lattice sites. We thus approximate: $J_{ij}s_is_j \rightarrow J_{ij}s_i\phi + J_{ij}s_j\phi$ where $\phi = \langle s_i \rangle$ is the average volume fraction of "B" particles. Now the sum in Eq. (1.63) can be evaluated since the sites are decoupled and where we approximate Z_μ by

$$Z_\mu = \left(1 + \exp\left[\phi J/T + (\mu - J/2)/T\right]\right)^N \qquad (1.65)$$

Since the interaction is translationally invariant in an infinite system with N sites, J_{ij} depends only on the *difference* between the two site indices. We therefore define, $J = \sum_j J_{ij}$, so that $J\phi$ is the mean interaction of a solute molecule with the solvent and the background of other solute molecules (the mean field). The grand potential is of course given by: $G = -T \log Z_\mu$, while the Helmholtz free energy (see Eq. (1.13)), $F = G + N\mu\phi$ is

$$F = -TN \log\left(1 + \exp\left[\phi J/T + (\mu - J/2)/T\right]\right) + N\mu\phi \qquad (1.66)$$

The chemical potential is determined from the constraint $\langle s_i \rangle = \phi$ yielding

$$e^{\mu/T} = \left(\frac{\phi}{1 - \phi}\right) \exp\left[-\tfrac{1}{2}J(2\phi - 1)/T\right] \qquad (1.67)$$

so that the Helmholtz free energy per site, $f = F/N$:

$$f = T\left[\phi \log\phi + (1 - \phi)\log(1 - \phi)\right] + \frac{J}{2}\phi(1 - \phi) \qquad (1.68)$$

Since there is one molecule per site (either an "A" or a "B"), we can also refer to f as the free energy per molecule.

This free energy can be compared with that of the nearly ideal gas, Eq. (1.27); in both cases the free energy consists of an entropy term — calculated for the noninteracting system — and an interaction term. In Eq. (1.68), the first term represents the entropy of the noninteracting system; it is just related to the number of ways of occupying a fraction ϕ sites by the "B" particles for the case where the occupancies of the different particles are uncorrelated. The second term is the interaction energy between a fraction ϕ of "B" particles and $(1 - \phi)$ "A" particles in the random mixing approximation. This procedure can be easily generalized to treat an arbitrary number of interacting components in the random mixing approximation.

Free Energy Minima of Two Coexisting Phases

We now consider the problem of minimization of the free energy of a system with a fixed total volume and neglect compressions of the molecules. In the two-phase region of the phase diagram, the system consists of two coexisting phases ($i = 1, 2$) with a volume fraction ϕ_i of "B" particles in each of the phases. In addition to the volume fractions, we have to specify the number of sites (*i.e.*, volume in an incompressible system) occupied by each macroscopic phase and we denote as N_i the number of sites in each phase. Since the total number of sites, N, in the incompressible system is fixed, N_1 and N_2 are constrained:

$$N_1 + N_2 = N \qquad (1.69)$$

Similarly, the average volume fraction of "B" particles in the system is constrained to equal ϕ:

$$N_1\phi_1 + N_2\phi_2 = N\phi \qquad (1.70)$$

where $N_i\phi_i$ are the total number of "B" particles in each phase.

We thus must consider the minimization of the total free energy of the two co-existing phases, with the two constraints, Eq. (1.69,1.70). We therefore construct the Gibbs free energy, G, for the entire system:

$$G = [N_1 f(\phi_1) + N_2 f(\phi_2)] - \mu[N_1\phi_1 + N_2\phi_2] + \Pi[N_1 + N_2]v_0 \qquad (1.71)$$

including the constraints on both the number of particles in each phase and the volumes. In Eq. (1.71), the first term in brackets comes from the free energy per molecule, f, defined in Eq. (1.68), and μ and Π are Lagrange multipliers that enforce the constraints. The chemical potential, μ, enforces conservation of "B" particles while the osmotic pressure, Π, enforces conservation of volume in this incompressible system. The factor v_0 in Eq. (1.71) represents a molecular volume, so that Π has the units of a pressure (energy per unit volume).

Minimizing G with respect to the variables ϕ_i and N_i implies that in equi-librium one has equality of the chemical potentials and osmotic pressure. We thus have:

$$\frac{\partial f(\phi_1)}{\partial \phi_1} = \frac{\partial f(\phi_2)}{\partial \phi_2} = \mu \qquad (1.72)$$

and

$$\Pi = \frac{\phi_1^2}{v_0}\left(\frac{\partial f(\phi_1)/\phi_1}{\partial \phi_1}\right) = \frac{\phi_2^2}{v_0}\left(\frac{\partial f(\phi_2)/\phi_2}{\partial \phi_2}\right) \qquad (1.73)$$

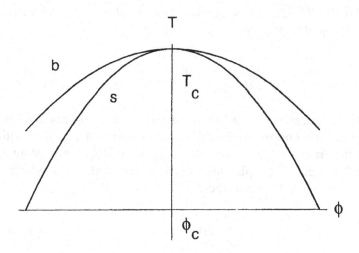

Figure 1.7 Schematic equilibrium coexistence curve or binodal (curve marked b) and spinodal curve (marked s, and defined in the text as the curve which is the limit of stability of the uniform phase) as a function of the volume fraction ϕ and temperature. In the entire region below the curve marked b, the system is no longer in a single phase.

One can regard Eqs. (1.72,1.73) as two algebraic equations that determine the values of ϕ_1 and ϕ_2 that coexist in equilibrium thus fixing the *coexistence curve* as a function of T and ϕ. Geometrically, these two conditions can be combined to yield a "common tangent" construction for the free energy so that

$$\mu = \frac{f(\phi_1) - f(\phi_2)}{\phi_1 - \phi_2} \tag{1.74}$$

This condition can be used to graphically solve for the two coexisting phases and hence to determine the coexistence curve that is schematically shown in Fig. 1.7. Often, however, it is easier to fix the volume fractions which are bounded ($0 \le \phi \le 1$) and determine the chemical potential from Eq. (1.72); the values of ϕ_1 and ϕ_2 that coexist are those which have the same values of μ and Π.

Instabilities and Critical Point for Phase Separation

The critical point is defined as the point (ϕ_c, T_c) on the equilibrium phase diagram where the compositions of the two coexisting phases become identical:

$\phi_1 = \phi_2 = \phi_c$. From Eq. (1.72), expanding $\phi_1 = \phi_c + \delta$ and $\phi_2 = \phi_c - \delta$, we see that for small δ, the equality of chemical potentials implies

$$\left[\frac{\partial^2 f}{\partial \phi^2}\right]_{\phi_c} = 0 \tag{1.75}$$

The critical point is a special case of a locus of points in the (ϕ, T) plane where the single-phase system is unstable — i.e., $f - \mu\phi$ is no longer a minimum with respect to ϕ and the critical temperature is the highest possible temperature at which this instability occurs. The spinodal curve determines this locus, which is given by the curve in the (ϕ, T) plane where

$$\frac{\partial^2 f}{\partial \phi^2} = 0 \tag{1.76}$$

At the critical point, (ϕ_c, T_c), the coexistence curve (or binodal) and spinodal curve meet. It is important to note that the spinodal is never physically reached in thermal equilibrium (except at the critical point itself) since it is precluded by the first-order phase separation described in the (ϕ, T) plane by the coexistence curve. While the coexistence curve separates the single phase and two-phase regions in thermal equilibrium, kinetic effects can allow the existence of metastable states — i.e., one can have the system below the binodal curve and still observe only a single phase for a rather long time. In contrast, the region below the spinodal curve is unstable and even small, thermal fluctuations will drive the system toward equilibrium. The spinodal line represents a line of large fluctuations in the concentration ϕ, since the free energy cost for fluctuations of ϕ away from its minimal values, as determined from Eq. (1.72), is proportional to $\partial^2 f / \partial \phi^2$; when this quantity is small, the fluctuation probability is large[1]. This line thus indicates a region where scattering should be large. However, reaching the spinodal involves a system that is not at thermodynamic equilibrium; this can be attained, for example, by a fast quench in temperature from the one-phase region to below the spinodal curve.

Because fluctuations become large at the critical point, the simple, mean-field theory used here breaks down. Large fluctuations mean that the approximation, $J_{ij}s_i s_j \to J_{ij}s_i \phi + J_{ij}s_j \phi$, used to simplify the partition function, Eq. (1.65), is no longer valid since the local value of the concentration is no longer approximately given by the average concentration. Even if these fluctuations are included as "corrections" to the mean-field approximation, the theory becomes quantitatively inaccurate near the critical point. A detailed theoretical treatment of these "critical phenomena" is outside the scope of this book (see for example

Ref. 24). However, analysis of both simple mean-field theories *plus* their fluctuation corrections includes most of the important physics and provides a guide to when one must include more sophisticated treatments very close to the critical point.

Landau Expansion of Free Energy near the Critical Point

For the mean-field free energies of either the lattice gas (Ising model) or imperfect gas, one would expect the free energies to have an analytic Taylor series expansion for the composition near the critical composition. This approach is known as a **Landau expansion** of the free energy (see Fig. 1.8 and Fig. 1.9) and plays a central role in the theory of phase transitions for both bulk systems and interfaces. Consider a system whose average composition is given by the critical composition, ϕ_c. The question is then whether the system remains stable in a single phase state with $\phi = \phi_c$ or whether the system phase separates into two coexisting phases whose relative volumes are such that the average composition is still ϕ_c. One assumes that when the composition is close to a given value, ϕ_0, one can expand the free energy per unit volume as an analytic function for small values of $\eta = \phi - \phi_0$. (This is not valid *very* close to the critical point where the mean-field approach breaks down[24].) The free energy per unit volume, f, can be written as a Taylor series expansion in η:

$$f = h\eta - \frac{\epsilon}{2}\eta^2 + \frac{b}{3}\eta^3 + \frac{c}{4}\eta^4 + \dots \qquad (1.77)$$

The linear term has a coefficient $h = (\partial f/\partial \phi)_{\phi_0}$. But this is just equal to the chemical potential, μ_0, at the volume fraction ϕ_0. Now, $\mu_0 = (\partial f/\partial \phi)_{\phi_0}$, so that if we set the chemical potential equal to this value, the grand potential, $g = f - \mu\eta = f - h\eta$, has no linear term in η:

$$g = -\frac{\epsilon}{2}\eta^2 + \frac{b}{3}\eta^3 + \frac{c}{4}\eta^4 + \dots \qquad (1.78)$$

Minimization of g with respect to η will then yield the compositions of the two coexisting phases and it is instructive to study this graphically in a plot of g versus η. When $b \neq 0$ we have a first-order transition with $\eta = 0$ for ϵ large and negative and η jumping to the value η_0 as ϵ is increased. In the two-phase region (where η is nonzero) the difference in the two coexisting compositions does *not* go smoothly to zero. On the other hand, when $b = 0$ (see Fig. 1.8) the two coexisting phases have compositions whose difference goes to zero at the critical point where $\eta \to 0$ and $\epsilon \to 0$, as can be seen by minimizing Eq. (1.78) with $b = 0$. The value of the composition ϕ_0 at which $b = 0$ when $\epsilon = 0$, is

g

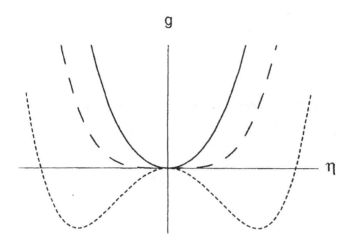

Figure 1.8 Free energy, g, versus order parameter, η, for a second-order phase transition; the solid curve is for $\epsilon < 0$, the dashed curve for $\epsilon = 0$, and the dotted curve is for $\epsilon > 0$.

denoted as the critical composition, ϕ_c. Thus at this special value, an expansion of the free energy around $\phi_0 = \phi_c$ for $\mu = \mu_c$, yields a free energy of the form of Eq. (1.78) but with $b = 0$. In addition, when the chemical potential is equal to the critical value μ_c (so that the average composition is equal to ϕ_c), minimization of g implies that $\eta_1 = -\eta_2$; this is consistent with the equality of the osmotic pressures in the two phases, so that the minimization of g determines the phase coexistence and hence the binodal.

Since Eq. (1.77,1.78) come from a series expansion of the free energy, this is consistent with the identification of the critical point with

$$\epsilon = \left(\frac{\partial^2 f}{\partial \phi^2} \right)_{\phi_c, T_c} = 0 \qquad (1.79a)$$

as mentioned previously. Since, in general, ϵ and b are functions of temperature (or the temperature scaled by the interaction strength), the locations of the critical values of ϕ_c and T_c are given by the two conditions: Eq. (1.79a) and

$$\left(\frac{\partial^3 f}{\partial \phi^3} \right)_{\phi_c, T_c} = 0 \qquad (1.79b)$$

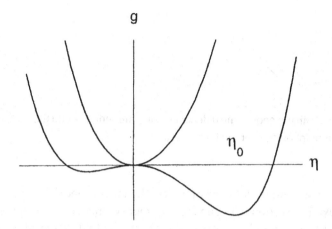

Figure 1.9 Free energy, g, versus order parameter, η, for a first-order phase transition where the value of η that minimizes g jumps from $\eta = 0$, for large negative values of ϵ, to a value η_0, for small values of ϵ.

An alternative way to demonstrate condition Eq. (1.79) is to note that for the free energy to be a minimum *at* the critical temperature when $T = T_c$ and $\epsilon = 0$ implies that $b = 0$ and $c > 0$.

Phase Separation in Polymer Solutions

The previous discussion, whose microscopic origins are in a lattice gas theory of mixtures, is applicable to mixtures of compact, molecules with comparable dimensions. In order to treat mixtures of long, chain, flexible polymers in small molecule solvents, the mean-field theory described above must be modified to take into account that the polymers and the solvent molecules are not of the same size and that the polymers are flexible macromolecules. The resulting[11,25] "Flory-Huggins" free energy per monomer, f_{FH}, is written

$$f_{FH} = T \left[\left(\frac{\phi}{N} \right) \log \left(\frac{\phi}{N} \right) + (1 - \phi) \log(1 - \phi) + \chi \phi (1 - \phi) \right] \quad (1.80)$$

A simple explanation of this expression is that it is a sum of the translational entropy of the polymer with N monomers per chain (first term in Eq. (1.80)) and the translational entropy of the solvent (second term in Eq. (1.80)). Since the polymers center of mass moves N units at once, its translational entropy is

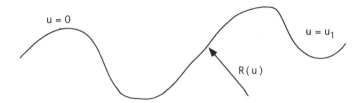

Figure 1.10 Space curve with the coordinate u describing the arc-length along the curve; $\vec{R}(u)$ is the position of a point on the curve at arc-length u.

smaller than that of the solvent by a factor of $1/N$. The term proportional to χ represents the attractive interaction between the polymer segments. If χ is large enough the system will undergo phase separation similar to that of the binary mixture described before. However, because of the factors of $1/N$, the critical concentration occurs at small values of the volume fraction if the molecular weight is large: $\phi_c = 1/(1+\sqrt{N})$. This is because the entropy is only comparable to the interaction energy at very small volume fractions due to the reduction by the factor of $1/N$. At the critical point, the value of χ (corresponding to T_c in the discussion of simple mixtures) depends only weakly on N for long chains and $\chi_c \approx 1/2$.

1.5   DIFFERENTIAL GEOMETRY OF SURFACES

The phenomenon of phase separation naturally gives rise to the presence of interfaces whose study is the focus of the chapters that follow. In this section, we discuss the mathematical definition[5] of an interface or surface (of zero thickness) and show how one can calculate the area and curvature of a general surface. These concepts are particularly useful in the statistical physics of surfaces, interfaces, and membranes since one often has terms in the energy that depend on the area (*e.g.*, surface tension) and/or curvature (*e.g.*, bending energy). Of course the physical problem is more complicated since often the interface shape and size is not known but is determined self-consistently by the system. Thus one often asks which surfaces have a given area or curvature. An additional complication is that the surfaces of interest are often not deterministic but rather stochastic; thermal fluctuations of the surface (surface entropy) must be taken into account as described in the chapters on fluctuations of surfaces and membranes.

Space Curves

A curve in space, as shown in Fig. 1.10, is described by a position vector $\vec{r} = \vec{R}(u)$ where u is a parameter that denotes the points along the curve in a simple, one-dimensional manner — e.g., u might label monomers in a polymer chain. The infinitesimal arc-length, du, is the distance along the contour of the chain between two points labeled by u_1 and u_2 in the limit that $u_1 \to u_2$. The *actual distance* in space between those points is denoted by s. For discrete points: $\Delta s = |\vec{R}(u_1) - \vec{R}(u_2)|$, which in the limit that $u_1 \to u_2$ is

$$ds = \left| \frac{\partial \vec{R}}{\partial u} \right| du \tag{1.81}$$

The unit tangent vector, $\hat{t}(u)$, is defined by

$$\hat{t} = \frac{d\vec{R}}{ds} = \frac{d\vec{R}/du}{ds/du} \tag{1.82}$$

One shows that $\hat{t}$ is indeed a unit vector by using Eq. (1.81). The rate of change of the tangent is related to the curvature, κ, and the normal, $\hat{n}_c$, by

$$\frac{d\hat{t}}{ds} = \kappa \hat{n}_c = \frac{d^2 \vec{R}}{ds^2} \tag{1.83}$$

Surfaces

Surfaces are described by either the parametric form $x = f(u, v), y = g(u, v)$, and $z = h(u, v)$, which determine a vector $\vec{r}(u, v)$, or by the implicit form $F(x, y, z) = 0$. A simple example of the parametric form is where u and v are equal to x and y respectively and the "height" $z = h(u, v) = h(x, y)$. This is called the **Monge parameterization** of a surface, as shown in Fig. 1.11. The position of the surface is given by

$$\vec{r} = (u, v, h(u, v)) = (x, y, h(x, y)) \tag{1.84}$$

Another example are spherical coordinates on a unit sphere so that u and v are the angles θ and ϕ respectively. On the surface one defines the *two* tangent vectors $\vec{r}_u = \partial \vec{r}/\partial u$ and $\vec{r}_v = \partial \vec{r}/\partial v$. These vectors are not necessarily unit vectors, nor are they necessarily orthogonal. The two vectors define a tangent plane: $d\vec{r} \cdot \hat{n} = 0$

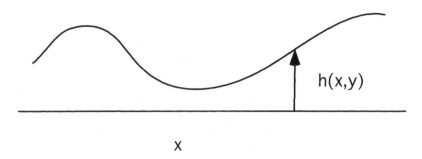

Figure 1.11 A surface above the xy plane; the z-coordinate of the surface is given by $h(x, y)$.

where $\hat{n}$ is the normal to the surface at "positions" (u, v). The normal is given by the cross product:

$$\hat{n} = \frac{\vec{r}_u \times \vec{r}_v}{|\vec{r}_u \times \vec{r}_v|} \tag{1.85}$$

If the implicit form of the surface is used, $F(x, y, z) = 0$, and in the Monge representation $F(x, y, z) = z - h(x, y) = 0$. For the general implicit form, one obtains the normal, by realizing that on the surface, where F is a constant, the total derivative of F is zero:

$$dF = d\vec{r} \cdot \nabla F = 0 \tag{1.86}$$

where $d\vec{r}$ is a vector connecting two points in the surface. Since $d\vec{r}$ is a vector that is tangent to a certain direction in the surface, Eq. (1.86) indicates that ∇F is orthogonal to this tangent; the normal vector is thus parallel to ∇F (with F evaluated on the surface). The unit normal is thus given by

$$\hat{n} = \frac{\nabla F}{|\nabla F|} \tag{1.87}$$

Metric of a Surface

Consider a surface defined by $\vec{r}(u, v)$. The distance in space, ds, between two points at u and $u + du$ and v and $v + dv$ is given by

$$(ds)^2 = (d\vec{r})^2 = (\vec{r}_u du + \vec{r}_v dv)^2 \tag{1.88}$$

One writes this as

$$(ds)^2 = E(du)^2 + 2F du\, dv + G(dv)^2 \tag{1.89}$$

where

$$E = \vec{r}_u^2; \qquad F = \vec{r}_u \cdot \vec{r}_v; \qquad G = \vec{r}_v^2 \tag{1.90}$$

The metric, g is defined by

$$g = EG - F^2 \tag{1.91}$$

It is a positive definite quantity since

$$g = (\vec{r}_u \times \vec{r}_v)^2 > 0 \tag{1.92}$$

The area of a parallelogram with sides $d\vec{r}_u$ and $d\vec{r}_v$ is the area element, dA, given by

$$dA = |\vec{r}_u \times \vec{r}_v|\, du\, dv = \sqrt{g}\, du\, dv \tag{1.93}$$

For the Monge gauge we write $\vec{r}_u = \vec{r}_x = (1, 0, h_x)$, $\vec{r}_v = \vec{r}_y = (0, 1, h_y)$ where the subscripts denote a derivative (e.g., $h_x = \partial h(x, y)/\partial x$). This allows us to compute the area element:

$$dA = dx\, dy \sqrt{1 + h_x^2 + h_y^2} \tag{1.94}$$

The normal is given by

$$\hat{n} = \frac{\hat{z} - h_x \hat{x} - h_y \hat{y}}{\sqrt{1 + h_x^2 + h_y^2}} \tag{1.95}$$

If the implicit form of the surface is used, $F(x, y, z) = 0$, the area element can be obtained by noting that the surface is the locus of points where $F = 0$. One then finds that the equation equivalent to Eq. (1.93) is

$$\int dA = \int d^3\vec{r}\, \delta(F)\, |\nabla F| \tag{1.96}$$

This can be verified explicitly for the case where $F(x, y, z) = z - h(x, y)$; performing the integration over the variable z yields Eq. (1.94) for the local area element. Since one can always *locally* define the surface using this parameterization, Eq. (1.96) for the area element is generally valid.

Surface Curvature

Consider a specific curve on a surface with distance, ds, given by Eq. (1.89). The normal curvature of this curve, is defined by

$$\kappa = \vec{r}^{\,\prime\prime} \cdot \hat{n} \tag{1.97}$$

where the prime signifies a derivative with respect to s (e.g., $\vec{r}^{\,\prime} = d\vec{r}/ds$) and $\hat{n}$ is the normal. Note that this is analogous, but not identical, to the definition of the curvature of a line in Eq. (1.83) since it chooses the relevant direction of the normal to the surface; curvature of the curve within the surface is not important in this regard. In special cases, one can choose $\hat{n} = \hat{n}_c$, but this is not true in general. Now, $\vec{r}^{\,\prime\prime}$ is given in terms of the derivatives in v and u as

$$\vec{r}^{\,\prime\prime} = u^{\prime\prime}\vec{r}_u + v^{\prime\prime}\vec{r}_v + (u^{\prime})^2\vec{r}_{uu} + (v^{\prime})^2\vec{r}_{vv} + 2(u^{\prime}v^{\prime})\vec{r}_{vu} \tag{1.98}$$

where as before the prime denotes differentiation with respect to the distance in space, s. Using Eqs. (1.89, 1.98) in Eq. (1.97) and remembering that $\hat{n} \cdot \vec{r}_u = \hat{n} \cdot \vec{r}_v = 0$ (from Eq. (1.85)), we find that the curvature is

$$\kappa = \frac{L(du)^2 + 2M\,du\,dv + N(dv)^2}{E(du)^2 + 2F\,du\,dv + G(dv)^2} \tag{1.99}$$

where

$$L = \hat{n} \cdot \vec{r}_{uu}; \qquad M = \hat{n} \cdot \vec{r}_{uv}; \qquad N = \hat{n} \cdot \vec{r}_{vv} \tag{1.100}$$

We note the relations $\hat{n} \cdot \vec{r}_u = \hat{n} \cdot \vec{r}_v = 0$, which arise from the orthogonality of the normal and the tangent vectors. Differentiating these relations, we define:

$$L = -\hat{n}_u \cdot \vec{r}_u; \qquad M = -\hat{n}_v \cdot \vec{r}_u = -\hat{n}_u \cdot \vec{r}_v; \qquad N = -\hat{n}_v \cdot \vec{r}_v \tag{1.101}$$

where the subscript on $\hat{n}$ denotes differentiation (e.g., $\hat{n}_u = d\hat{n}/du$). From this relation, one can see that Eq. (1.99) for the curvature can also be written as

$$\kappa = -\frac{d\vec{r} \cdot d\hat{n}}{d\vec{r} \cdot d\vec{r}} \tag{1.102}$$

For each direction in the surface, $d\vec{r}$, there is thus a value of the curvature.

This form for the curvature for a curve on the surface can also be derived from the implicit form of the surface, $F(x, y, z) = 0$ by considering the change in the normal vector defined from Eq. (1.87) as one proceeds along the surface. Thus, if one moves along the surface a distance $d\vec{r}$, the normal, $\hat{n}$, changes by an amount:

$$d\hat{n} = d\vec{r} \cdot \mathbf{Q} \qquad (1.103)$$

where $\mathbf{Q}$ is a tensor whose elements in Cartesian coordinates are given by differentiating Eq. (1.87):

$$Q_{ij} = \frac{1}{\Upsilon} \left[F_{ij} - \frac{F_i \Upsilon_j}{\Upsilon} \right] \qquad (1.104)$$

where $\Upsilon = |\nabla F|$ and $F_i = \partial F/\partial r_i$, where $\vec{r} = (x, y, z)$, with a similar notation for Υ_i. By taking the dot product of Eq. (1.103) with $d\vec{r}$ one obtains an expression similar to Eq. (1.102) for the curvature along a given direction. Comparison of these two expressions shows that the curvature, κ, is proportional to the trace of the tensor $\mathbf{Q}$. Thus, the curvature is associated with the change in the normal as one moves along the surface. Since both the normal and the direction along the surface are vectors, the curvature is, in general, a tensor quantity.

Invariants of the Curvature Tensor: Mean and Gaussian Curvatures

It is useful to describe the curvature tensor by its invariants, since these quantities do not change if one rotates the coordinate system used to describe the surface; the invariants are intrinsic properties of the surface. For the implicit representation of the surface, $F(x, y, z) = 0$, the curvature along a general direction is related to the tensor, $\mathbf{Q}$, defined in Eq. (1.104). In three dimensions, $\mathbf{Q}$ is a 3×3 matrix with three eigenvalues. One can show by explicit calculation using Eq. (1.104) with $\Upsilon = |\nabla F|$, that the determinant of $\mathbf{Q}$ and one eigenvalue are zero. The remaining two eigenvalues are the two principal curvatures of the surface; these can be calculated for any F directly from $\mathbf{Q}$.

The three-dimensional tensor $\mathbf{Q}$ has three invariants under similarity transformations (which include rotations): its trace, the sum of the principal minors (*i.e.*, the three minors formed by crossing out the rows and columns of the diagonal elements), and its determinant (for a proof see the following example). Explicit calculation shows that the determinant of $\mathbf{Q}$ is zero. Two of the eigenvalues of $\mathbf{Q}$ have dimensions of an inverse length and are known as the principal curvatures (one eigenvalue is zero). The eigenvectors corresponding to the nonzero eigenvalues are known as the principal directions of the surface; along these directions, the curvature tensor is diagonal. The trace, which is the sum of the eigenvalues (*i.e.,* principal curvatures), is twice the mean curvature, H; both quantities have the dimensions of an inverse length. The other invariant of the tensor $\mathbf{Q}$ is the sum of the principal minors, which has the dimensions of an inverse length squared and is termed the Gaussian curvature, K, which is equal to the product of the two principal curvatures (eigenvalues of $\mathbf{Q}$). Using Eq. (1.104) we find an expression for the mean curvature:

$$H = \frac{1}{2\Upsilon^3} \left[F_{xx}(F_y^2 + F_z^2) - 2F_x F_y F_{xy} + \text{Perm} \right] \qquad (1.105)$$

where the term **Perm** indicates that one should consider two additional permutations of each term — one where $(x, y, z) \to (z, x, y)$ and another with $(x, y, z) \to (y, z, x)$ — and where

$$\Upsilon = \sqrt{F_x^2 + F_y^2 + F_z^2} \qquad (1.106)$$

The Gaussian curvature is given by

$$K = \frac{1}{\Upsilon^4} \left[F_{xx} F_{yy} F_z^2 - F_{xy}^2 F_z^2 + 2F_{xz} F_x (F_y F_{yz} - F_z F_{yy}) + \text{Perm} \right] \quad (1.107)$$

In the case where F is described by the Monge parameterization $F = z - h(x, y)$, these expressions simplify considerably as discussed below and in the problems at the end of this chapter.

Example: Invariants of a Tensor

Show that for a general 3×3 matrix, $\mathbf{A}$, the trace, sum of the principal minors (*i.e.,* the three minors formed by crossing out the rows and columns of the diagonal elements), and the determinant are invariant under similarity transformations.

Consider the characteristic polynomial of $\mathbf{A}$ given by

$$P(\lambda) = \det(\mathbf{A} - \lambda \mathbf{I}) \qquad (1.108)$$

where $\mathbf{I}$ is the unit matrix. We first show[26] that $P(\lambda)$ is invariant under a similarity transformation where $\mathbf{A} \to \mathbf{C}^{-1}\mathbf{A}\mathbf{C}$. Writing the determinant as $\det \mathbf{A} = |\mathbf{A}|$, we write

$$P'(\lambda) = |\mathbf{C}^{-1}\mathbf{A}\mathbf{C} - \lambda \mathbf{I}| = |\mathbf{C}^{-1}\mathbf{A}\mathbf{C} - \lambda \mathbf{C}^{-1}\mathbf{C}|$$

$$= |\mathbf{C}^{-1}(\mathbf{A} - \lambda \mathbf{I})\mathbf{C}| \qquad (1.109)$$

Using the fact that $|\mathbf{XY}| = |\mathbf{X}|\,|\mathbf{Y}|$, we find that

$$P'(\lambda) = |\mathbf{C}^{-1}|\,|\mathbf{A} - \lambda\mathbf{I}|\,|\mathbf{C}| \qquad (1.110)$$

Finally, using the fact that $|\mathbf{C}^{-1}| = 1/|\mathbf{C}|$, we see that $P'(\lambda) = P(\lambda)$; the characteristic polynomial is invariant under rotations or other similarity transformations.

Now, an explicit calculation of the characteristic polynomial of a 3×3 matrix, $\mathbf{A}$ with elements A_{ij} yields

$$P(\lambda) = |\mathbf{A}| - \lambda \sum_{i=1,2,3} M_i + \lambda^2 \sum_{i=1,2,3} A_{ii} - \lambda^3 \qquad (1.111)$$

where M_i is the minor formed by crossing out the rows and columns of each diagonal element. Thus, the term proportional to λ^2 is the trace and the term proportional to λ is the sum of the principal minors. Since $P(\lambda)$ is invariant under similarity transformations and since λ is arbitrary, each term in Eq. (1.111) must be separately invariant under similarity transformations. Thus, the determinant, trace, and sum of principal minors are all invariants of the 3×3 matrix. ∎

Parametric Representation: Mean and Gaussian Curvatures

We now derive expressions for the mean and Gaussian curvatures using the parametric form of the surface, $x = f(u, v)$, $y = g(u, v)$, and $z = h(u, v)$. We show that the two principal curvatures are the extrema of all possible surface curvatures at a given point. Consider a unit vector in the surface, $\hat{a} = \ell\vec{r}_u + m\vec{r}_v$. Since $a^2 = E\ell^2 + 2F\ell m + Gm^2 = 1$, the curvature along this unit vector is given by Eq. (1.99) as

$$\kappa = L\ell^2 + 2M\ell m + Nm^2 \qquad (1.112)$$

With the constraint of a unit vector, we want to find the curves — i.e., the values of (ℓ, m) that maximize or minimize the curvature. We introduce a Lagrange multiplier, λ and seek the extrema of $\tilde{\kappa}$

$$\tilde{\kappa} = L\ell^2 + 2M\ell m + Nm^2 - \lambda(E\ell^2 + 2F\ell m + Gm^2) \qquad (1.113)$$

Setting $\partial\tilde{\kappa}/\partial\ell = \partial\tilde{\kappa}/\partial m = 0$, we get two equations. The multiplier, λ can be shown to obey $\lambda = \kappa$, with κ given by Eq. (1.112). By finding the values of the variables ℓ and m where $\tilde{\kappa}$ is an extremum and putting those values into

Eq. (1.112) we can find the extremal values of the curvature κ. These values are determined by the quadratic equation:

$$\kappa^2(EG - F^2) - \kappa(EN + GL - 2FM) + (LN - M^2) = 0 \qquad (1.114)$$

There are two roots of this equation: κ_a and κ_b, which define the mean curvature, $H = \frac{1}{2}(\kappa_a + \kappa_b)$, and the Gaussian curvature, $K = \kappa_a\kappa_b$, through

$$H = \frac{EN + GL - 2FM}{2(EG - F^2)} \qquad (1.115)$$

$$K = \frac{LN - M^2}{EG - F^2} \qquad (1.116)$$

Another way to interpret the procedure described is that the curvature along an arbitrary direction, in the surface $\hat{a} = \ell\vec{r}_u + m\vec{r}_v$, is a quadratic function of the values of ℓ and m. Diagonalizing this quadratic form, subject to the constraint that $\hat{a}$ is a unit vector, is mathematically equivalent to the minimization/maximization of κ. Thus, the extremal curvatures, κ_a and κ_b are determined by the extremal values of ℓ and m, which we denote as ℓ^* and m^*. For directions on the surface close to these extremal values, the expansion of the curvature as a function of $(\ell - \ell^*)$ and $(m - m^*)$ has no linear terms since ℓ^* and m^* are extremal; in this sense, the local curvature is a diagonalized quadratic form.

Principal Directions

The two directions whose curvatures are extrema are known as the principal directions. These two directions can be shown to be orthogonal if one eliminates the multiplier λ from the equations $\partial\tilde{\kappa}/\partial\ell = \partial\tilde{\kappa}/\partial m = 0$ to get

$$(EM - FL)\ell^2 + (EN - GL)\ell m + (FN - GM)m^2 = 0 \qquad (1.117)$$

which has two solutions for the ratio (ℓ/m) with curvatures given by κ_a and κ_b (see Eqs. (1.115,1.116)). These two solutions of Eq. (1.117), $(\ell/m)_1 = \mu_1$ and $(\ell/m)_2 = \mu_2$ have a product of roots and sum of roots that obey

$$\mu_1 + \mu_2 = -\frac{EN - GL}{EM - FL} \qquad (1.118)$$

$$\mu_1\mu_2 = \frac{FN - GM}{EM - FL} \qquad (1.119)$$

(These relations come from the theory of quadratic equations.) Now, the two directions have unit vectors $\hat{a}_1 = m_1(\mu_1\vec{r}_u+\vec{r}_v)$ and $\hat{a}_2 = m_2(\mu_2\vec{r}_u+\vec{r}_v)$. The orthogonality condition, $\hat{a}_1 \cdot \hat{a}_2 = 0$ can be shown to be obeyed when Eqs. (1.118,1.119) apply. Thus the principal directions on the surface are orthogonal and define the extrema of the curvatures. (The orthogonality condition requires that $F = 0$.) If, in addition, these curves are used to parametrically define the surface — i.e., if they define the parameters u and v, then the parametric lines $du = 0$ and $dv = 0$ mean that either $\ell = 0$ or $m = 0$. From Eq. (1.117), this requirement along with the orthogonality condition, $F = 0$, also implies that $M = 0$. Along the principal directions the two curvatures therefore obey $\kappa_a = L/E$ and $\kappa_b = N/G$.

Example: Curvatures in the Monge Representation

If a surface is represented by $z = h(x, y)$, what are the mean and Gaussian curvatures in terms of the derivatives of h with respect to x and y? What are the limiting forms of these expressions when the slopes are small and terms in $h_x = \partial h(x, y)/\partial x$ and h_y can be neglected in the final formulas?

Using Eqs. (1.115,1.116) and the definitions in Eqs. (1.90,1.100) the curvatures are computed using $\vec{r} = x\hat{x} + y\hat{y} + h(x, y)\hat{z}$. The normal is given by Eq. (1.95) and $E = r_x^2 = 1 + h_x^2$, $G = 1 + h_y^2$, and $F = h_x h_y$. Similarly, $L = h_{xx}/\Upsilon$, $M = h_{xy}/\Upsilon$, and $N = h_{yy}/\Upsilon$, where $\Upsilon = \sqrt{1 + h_x^2 + h_y^2}$. The curvatures are thus written:

$$H = \frac{(1 + h_x^2)h_{yy} + (1 + h_y^2)h_{xx} - 2h_x h_y h_{xy}}{2\sqrt{(1 + h_x^2 + h_y^2)^3}} \tag{1.120}$$

$$K = \frac{h_{xx}h_{yy} - h_{xy}^2}{(1 + h_x^2 + h_y^2)^2} \tag{1.121}$$

In the limit of a nearly flat surface, $h_x \ll 1, h_y \ll 1$; the mean and Gaussian curvature can be approximately written as

$$H \approx \tfrac{1}{2}(h_{xx} + h_{yy}) \tag{1.122}$$

$$K \approx h_{xx}h_{yy} - h_{xy}^2 \tag{1.123}$$

∎

(u_0, v_0)

Figure 1.12 Curvatures about the point (u_0, v_0) for the case where the curvatures are positive. When one curvature is positive and the other negative, one gets a saddle-shaped region.

Physical Meaning of Curvatures

Along these principal directions, one can expand the position vector in a Taylor series about a given position (u_0, v_0). Forming the projection of the position on the normal vector $(\vec{r} \cdot \hat{n})$ to give the "height", z, of the surface (see Fig. 1.12), the first-order terms in $u - u_0$ and $v - v_0$ vanish because of the orthogonality of the normal and the tangent vectors. In addition, the use of the principal directions as parametric curves implies $M = 0$ as stated earlier, so $\hat{n} \cdot \vec{r}_{uv} = 0$ and there are no terms in $(u - u_0)(v - v_0)$ in the expansion. Thus, the equation of the surface is locally given by:

$$z = \hat{n} \cdot \vec{r}(u, v) = \vec{r}(u_0, v_0) \cdot \hat{n} + \tfrac{1}{2}(u - u_0)^2(\hat{n} \cdot \vec{r}_{uu}) + \tfrac{1}{2}(v - v_0)^2(\hat{n} \cdot \vec{r}_{vv}) \quad (1.124)$$

Along the principal directions, the quadratic form for the local position of the surface is thus diagonal. Writing $z_0 = \vec{r}(u_0, v_0) \cdot \hat{n}$ and noting that $M = 0$ implies that the term proportional to $\vec{r}_{uv} = 0$, we have

$$z = z_0 + \tfrac{1}{2}L(u - u_0)^2 + \tfrac{1}{2}N(v - v_0)^2 \quad (1.125)$$

Converting to rectangular coordinates (we have already chosen the $\hat{z}$ directions along the normal), $\vec{x} = u\,\vec{r}_u$ and $\vec{y} = v\,\vec{r}_v$. Recalling the definitions of E and G and the values of the curvature along the principal directions, we have

$$z = z_0 + \tfrac{1}{2}\kappa_a(x - x_0)^2 + \tfrac{1}{2}\kappa_b(y - y_0)^2 \quad (1.126)$$

Thus, the principal curvatures are related to the *local* expansion of the surface near a given point. In a general direction, the normal curvature is a combination of κ_a and κ_b.

Parallel Surfaces

Consider a locally flat surface (*e.g.*, in the Monge gauge $|\nabla h| \ll 1$) and imagine translating that surface along the normal direction, a distance $\delta > 0$ to obtain a parallel surface whose location,

$$\vec{r}\,' = \vec{r}(u, v) - \hat{n}(u, v)\,\delta \qquad (1.127)$$

The relation of the area element (dA'), mean curvature (H'), and Gaussian curvature (K') on the parallel surface compared with the original surface with area dA, and curvatures H and K is given by

$$dA' = dA\left[1 + 2H\,\delta + K\,\delta^2\right] \qquad (1.128)$$

$$H' = \frac{H + K\,\delta}{1 + 2H\,\delta + K\,\delta^2} \qquad (1.129)$$

$$K' = \frac{K}{1 + 2H\,\delta + K\,\delta^2} \qquad (1.130)$$

and the normals are of course equal to within a sign. These relations are important in understanding the properties of surfaces of finite thickness. (The proof is straightforward, but lengthy and simplifies if one chooses the lines of curvature as parametric curves.)

Example: Minimal Surfaces — Euler-Lagrange Equations

Show that a surface of minimal area has zero curvature using the Monge representation. What are the relations between the two curvatures and what is the Gaussian curvature?

In the Monge representation of the surface, we have from Eq. (1.94) that, the total area A for a surface described by a height $h(x, y)$ above the xy plane is

$$A = \int dx\,dy\,\sqrt{1 + h_x^2 + h_y^2} \qquad (1.131)$$

where the integral is over a *fixed* area of the surface projection in the xy plane. The area is a function of the *derivatives* of $h(x, y)$ and to minimize A with respect to all possible surface shapes (*i.e.*, all possible functions $h(x, y)$), one uses the Euler-Lagrange equations of the calculus of variations[27] that determine the function which minimizes a given integral of that function. For example, if the integral $I = \int f[\psi(\vec{r}), \nabla\psi(\vec{r})]$, depends on both a function, $\psi(\vec{r})$, and its first derivatives, the Euler-Lagrange equations which determine the minimal $\psi(\vec{r})$ are

$$\frac{\delta I}{\delta \psi(\vec{r})} = \frac{\partial f}{\partial \psi(\vec{r})} - \nabla \cdot \frac{\partial f}{\partial \nabla \psi(\vec{r})} = 0 \qquad (1.132)$$

This is true for the case where the function $\psi(\vec{r})$ is fixed at the boundaries of the integration; if $\psi(\vec{r})$ is free at a given boundary, one can show[27] that the minimization implies the boundary condition:

$$\hat{n} \cdot \frac{\partial f}{\partial \nabla \psi(\vec{r})} = 0 \qquad (1.133)$$

where $\hat{n}$ is the normal to the boundary and the derivative is evaluated at the free boundary.

In our case, $\psi = h(x, y)$ and $f = \sqrt{1 + h_x^2 + h_y^2}$ and is only a function of the derivatives. Then Eq. (1.132) states that

$$\frac{\partial}{\partial x} \frac{h_x}{\sqrt{1 + h_x^2 + h_y^2}} + \frac{\partial}{\partial y} \frac{h_y}{\sqrt{1 + h_x^2 + h_y^2}} = 0 \qquad (1.134)$$

This expression can be written as

$$\frac{(1 + h_x^2)h_{yy} + (1 + h_y^2)h_{xx} - 2h_x h_y h_{xy}}{\sqrt{(1 + h_x^2 + h_y^2)^3}} = 0 \qquad (1.135)$$

From Eq. (1.120) we see that this is equivalent to the vanishing of the average curvature, H in the Monge representation. Thus, the minimal area surface is one that has zero mean curvature; the two curvatures are equal and opposite, implying that the Gaussian curvature is *negative*. These surfaces are therefore saddle-shaped and do not have the topology of a sphere or an ellipsoid (which have positive Gaussian curvature).   ∎

1.6   REVIEW OF HYDRODYNAMICS

Although most of the topics treated here will involve only the static properties of interfaces and membranes, many interesting and technologically important problems involve the motion of interfaces (*e.g.,* wetting of solid surfaces by fluids). For these problems, the microscopic properties of the interfaces are lumped into a few parameters such as the interfacial tension and the fluid is described by its local, time-dependent density, $\rho(\vec{r}, t)$, and velocity $\vec{v}(\vec{r}, t)$. Motion of the interface is coupled to motion of the bulk fluid and this coupling is expressed by the fact that the time derivative of the interface position is equivalent to the fluid velocity at the interface. Interfaces and surfaces arise naturally in hydrodynamic problems since velocity gradients that give rise to dissipation are often due to the presence of solid boundaries where the fluid "sticks" (so-called **no-slip** boundary conditions) and the velocity vanishes. Here we highlight the major concepts and equations that are useful in studying interface motion in fluids[6]. Several examples are presented in Chapter 4, where the dynamics of wetting is discussed.

Conservation of Matter

The conservation of matter is expressed by the continuity equation

$$\frac{\partial \rho}{\partial t} + \nabla \cdot (\rho \vec{v}) = 0 \tag{1.136}$$

where ρ is the density and $\vec{v}(\vec{r}, t)$ is the velocity. For incompressible flow, ρ is a constant so that $\partial \rho / \partial t = 0$ and the continuity equation implies

$$\nabla \cdot \vec{v} = 0 \tag{1.137}$$

Total Derivative

In discussing forces and motion of particles one uses Newton's law, which states that the acceleration of a particle is proportional to the force acting on the particle. If we denote $d\vec{v}/dt$ to mean the rate of change of the velocity of a given fluid particle as it moves in space, Newton's equation becomes

$$\rho \frac{d\vec{v}}{dt} = \vec{f} \tag{1.138}$$

where $\vec{f}$ is the force per unit volume. However, in discussing continuum fluids, it is more convenient to think of the velocity field at a fixed point in space as the particles move through the fluid via this point. Thus, the differential change in the velocity, $\vec{v}$, in a time dt, involves the difference between $\vec{v}(\vec{r} + \vec{v}dt,\, t + dt)$ and $\vec{v}(\vec{r},\, t)$ since there is an explicit time rate of change of the velocity of a given particle which is at the point $\vec{r}$, as well as a contribution from the difference in velocities as the particle travels between two points a distance $d\vec{r}$ apart during the time dt:

$$d\vec{v} = \frac{\partial \vec{v}}{\partial t} dt + (d\vec{r} \cdot \nabla)\vec{v} \qquad (1.139)$$

Thus, the acceleration, $d\vec{v}/dt$ is written:

$$\frac{d\vec{v}}{dt} = \frac{\partial \vec{v}}{\partial t} + (\vec{v} \cdot \nabla)\vec{v} \qquad (1.140)$$

This concept is generalized in the definition of the total derivative, where

$$\frac{d}{dt} = \frac{\partial}{\partial t} + (\vec{v} \cdot \nabla) \qquad (1.141)$$

Newton's Law

Consider a volume of fluid. The sum of all the forces on all of the particles within this volume is the total force, $\vec{F}$, which obeys Newton's law:

$$\vec{F} = \int dV \rho \frac{d\vec{v}}{dt} \qquad (1.142)$$

Here, $d\vec{v}/dt$ is the local acceleration, ρ is the local density, and V is the volume. Next, consider a small volume with forces on the surface that are given by a stress tensor, $\mathbf{\Pi}$, where the surface force, $\vec{F}_s$, is given by

$$\vec{F}_s = \int \mathbf{\Pi} \cdot dS = \int (\nabla \cdot \mathbf{\Pi})\, dV \qquad (1.143)$$

and where the divergence theorem has been used to convert the surface integral to a volume integral. The total force per unit volume is then

$$(\nabla \cdot \mathbf{\Pi}) + \rho \vec{f} = \rho \frac{d\vec{v}}{dt} \qquad (1.144)$$

where $\rho\vec{f}$ is an *external* body force per unit volume (like gravity or an electric field), which is linear in the density. The divergence of the stress tensor accounts for the *internal* forces between molecules that are present when the velocity is not constant. The stress tensor is written phenomenologically as

$$\mathbf{\Pi} = -p\mathbf{I} + \eta\nabla\vec{v} \tag{1.145}$$

where p is the pressure, η is the viscosity, and $\mathbf{I}$ is the unit tensor (see Ref. 6 for a more precise definition in terms of the symmetric components of this tensor). The pressure is related to force per unit volume acting within the system. The divergence of the stress tensor is

$$\nabla \cdot \mathbf{\Pi} = -\nabla p + \eta\nabla^2\vec{v} \tag{1.146}$$

So the force is related to the gradient of the pressure and the second derivative of the velocity. Combining Eqs. (1.140,1.144,1.146) results in the famous Navier-Stokes equation for the balance of forces in a fluid:

$$\rho\left[\frac{\partial\vec{v}}{\partial t} + (\vec{v}\cdot\nabla)\vec{v}\right] = -\nabla p + \eta\nabla^2\vec{v} + \rho\vec{f} \tag{1.147}$$

For the incompressible fluid where ρ is a constant, the variables to be solved for are $\vec{v}$ (3 components) and p (1 variable). There are four equations to use: Eq. (1.137) and the three components of Eq. (1.147). For a compressible fluid, an equation of state is needed to relate changes in the density to the pressure. Most simple fluids are well approximated as incompressible for purposes of studies of wetting and surface properties.

Boundary Conditions

To solve the equations for the pressure and velocity of the fluid, one must specify boundary conditions. Usually one assumes that the fluid sticks to a solid wall, so that $\vec{v}(\vec{r}, t) = 0$ when $\vec{r}$ is on the solid surface. (This no-slip boundary condition may not be completely accurate at microscopic length scales.) The other boundary condition that is often important is that flows at infinity are unperturbed by the boundaries. Finally, at surfaces or interfaces, there is continuity of the normal and tangential forces. The force on a surface is related to the normal component of the stress tensor, Eq. (1.143); the ith component of the force per unit area, $\vec{f}_s = \partial\vec{F}_s/\partial S$, obeys

$$f_{s,i} = \Pi_{ik}n_k = -p\,n_i + \eta\frac{\partial v_k}{\partial x_i}n_k \tag{1.148}$$

where one sums over the repeated index (k). The term proportional to the viscosity is the frictional force on the surface. At a free surface where there are no tangential or normal forces:

$$\Pi_{ik} \, n_k = 0 \qquad (1.149)$$

Finally, there is the kinematic boundary condition that is just a restatement of the fact that the fluid velocity at the interface is equivalent to the *total* time derivative of the interface position. Expressing the total derivative using the convective terms discussed previously, we find that an interface whose position is described by a vector $\vec{R}$ obeys

$$\frac{\partial \vec{R}}{\partial t} + (\vec{v}_s \cdot \nabla) \, \vec{R} = \vec{v}_s \qquad (1.150)$$

where v_s is the velocity at the surface or interface. This equation is used to relate the hydrodynamic flow in the fluid to the motion of the position of the interface.

Reynolds Number (Re)

The Reynolds number measures the relative importance of inertial terms to friction (viscosity) terms. It is defined by $Re = (\ell v \rho)/\eta$, where ℓ is a typical length scale. When $Re \ll 1$ one is describing a high viscosity flow (laminar flow). When $Re \gg 1$ the inertial terms are dominant (potential flow). An ideal fluid has $\eta = 0$ and only shows potential flow. Near walls, which are important for surface and interface problems, the velocity drops to zero so the important regime is that of low Re.

Laminar Flows in Steady State

When the system is nearly in steady state and the velocities are small, one can neglect $(\vec{v} \cdot \nabla)\vec{v}$ terms in Eq. (1.147) and assume incompressible flow; this is called "creeping flow". The steady-state (where the inertial term $\partial \vec{v}/\partial t$ can be dropped) velocity field obeys in this limit:

$$\eta \nabla^2 \vec{v} = \nabla p \qquad (1.151)$$

Example: Poiseuille Flow Through a Channel

In equilibrium, the pressure in a system is uniform; there is no net force. The existence of pressure gradients in a system implies that the system is not in equilibrium — *i.e.*, there is a nonzero velocity (see Fig. 1.13). One can thus say that pressure gradients induce flow. A particularly simple example

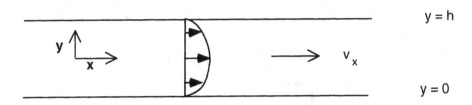

Figure 1.13 Flow through a channel induced by a pressure gradient. The velocity, $\vec{v}$, is zero at the walls located at $y = 0, h$.

of this phenomenon is given by the flow field in the region between two solid plates with no-slip boundary conditions. Here, symmetry dictates that the velocity field has a component only in the $\hat{x}$ direction, which can vary spatially in the $\hat{y}$ direction: $v_y = v_z = 0$, and $v_x = v_x(y)$. The pressure varies only in the $\hat{x}$ direction, $p = p(x)$. In steady state where the velocity fields are small, Eq. (1.151) becomes

$$\eta \frac{\partial^2 v_x}{\partial y^2} = \frac{\partial p}{\partial x} \qquad (1.152)$$

The right-hand side is a function of x only and the left-hand side a function of y, so Eq. (1.152) implies that each side is a constant. The constant is determined from the no-slip boundary condition $v_x = 0$ at $y = 0$ and $y = h$. The solution for the velocity as a function of the pressure gradient is

$$v_x = -\frac{1}{8\eta} \frac{\partial p}{\partial x} \left[h^2 - 4(y - \tfrac{1}{2}h)^2 \right] \qquad (1.153)$$

The velocity profile is *quadratic* as a function of y. The velocity, v_x is positive if the pressure gradient is negative; this expresses the fact that the material will flow in the positive $\hat{x}$ direction if the pressure *decreases* as one moves from lower to higher values of x. ∎

Example: Shear Flow

To obtain a shear flow, one considers a flow between two surfaces where one surface is fixed while the other moves with a given velocity as shown in Fig. 1.14. There is therefore a velocity gradient imposed by the boundary conditions at the walls. Here we take the boundary conditions on the velocity that set $v(y = 0) = 0$, and $v(y = h) = u$ where u is a constant. The pressure

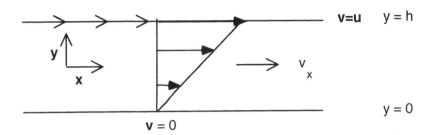

Figure 1.14 Shear flow where the velocity is zero at the wall at $y = 0$ and is fixed at a value of $\vec{u}$ at the wall at $y = h$.

gradients are zero and $\eta \frac{\partial^2 v_x}{\partial y^2} = 0$. The solution to Eq. (1.152) is a *linear* flow profile with $v_x = yu/h$.   ∎

Example: Gravity Driven Flow with a Free Surface

Here the body force, $\vec{f} = \rho \vec{g}$, is the gravitational force. With the reference system shown in Fig. 1.15, the component of $\vec{g}$ along the plane (which defines the $\hat{x}$-direction) is $g \sin \alpha$ so that the Navier-Stokes equation reads:

$$\eta \frac{\partial^2 v_x}{\partial y^2} + \rho g \sin \alpha = 0 \tag{1.154}$$

In the $\hat{y}$-direction there is no flow, $v_y = 0$ and the pressure gradient and gravity force balance:

$$\frac{\partial p}{\partial y} = -\rho g \cos \alpha \tag{1.155}$$

The boundary conditions are $v_x = 0$ at $y = 0$ and the free surface has no stresses:

$$\Pi_{xy} = \eta \frac{\partial v_x}{\partial y} = 0$$

at $y = h$. Also the yy component of the stress tensor is continuous, so the pressure equals the atmospheric pressure p_0 at $y = h$. The solutions for the velocity and pressure are

$$v_x = \frac{\rho g \sin \alpha}{2\eta} \, y(2h - y) \tag{1.156}$$

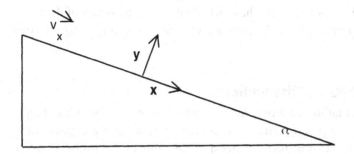

Figure 1.15 Flow down an inclined plane with angle of inclination α.

$$p = p_0 + \rho g \cos \alpha (h - y) \tag{1.157}$$

This treatment does not take the interactions of the fluid with the substrate into account. Wetting considerations lead to an unusual surface profile and to hydrodynamic instabilities as discussed in Chapter 4 in the section on wetting dynamics.   ■

1.7   PROBLEMS

1. Nearly Ideal Gas

Consider the nearly ideal gas with attractive interactions in the low-density expansion discussed previously. For large enough attractions, the system will undergo a gas-liquid phase transition. The virial expansion for the free energy per unit volume, f, as a function of density, n, to third order is

$$f = T \left[n \left(\log n v_0 - 1 \right) + \tfrac{1}{2} a n^2 + \tfrac{1}{3} b n^3 \right] \tag{1.158}$$

The coefficient a has been discussed in the text and the coefficient b can be obtained by higher order expansions of the free energy. Treating a as a parameter that will be varied (*e.g.*, through temperature changes, or through chemical changes that change the microscopic interactions) and $b > 0$ as a fixed, known quantity, compute the critical density, $n = n_c$, and the critical value of $a = a_c$ for phase separation. Now, expand the free energy near the critical point — *i.e.*, for values of $n \approx n_c$ and $a \approx a_c$. Using this expansion: (i) Show that in such an expansion, the third-order term vanishes. (ii) What are the densities of the coexisting gas and liquid phases? (iii) What is the chemical potential at equilibrium of these phases?

2. Gaussian Probability Distributions

Consider a fluctuating variable, $h(x, y)$, which may represent the height of an interface above a given plane to be described by a Gaussian probability distribution for its Fourier transform, as in Eq. (1.48). What is the probability that the variable $h(x = 0, y = 0) = h_0$? What is the probability that $h(x_0, y_0) = 0$? What is the joint probability that $h(x = 0, y = 0) = 0$ and that $|\nabla h(0, 0)| = \beta$? [*Note*: All of these probabilities can only be expressed as sums over $\vec{q}$ of functions of $G(\vec{q})$ until $G(\vec{q})$ is specified.]

3. Scattering from Spherical Particles

When scattering from noninteracting finite sized, colloidal particles, the scattering intensity reduces to the form factor for scattering from each individual particle:

$$I(\vec{q}) = I_0 \langle n(\vec{q})n(-\vec{q}) \rangle$$

where $n(\vec{q})$ is the Fourier transform of the density of each particle. Derive an expression for the form factor, $I(\vec{q})$ for scattering from a noninteracting system of spherical particles with radius R.

Derive an expression for the scattering, assuming that the scattering intensity only comes from the surface of the particle and not the entire bulk, which is contrast matched to the solvent and therefore shows no scattering.

4. Correlation Function and Scattering from a Polymer

Derive an expression for the Fourier transform of the density-density correlation function and hence the scattering from a system of dilute, and hence noninteracting polymers. The chains are described by the position $\vec{R}(s)$ of monomer s along the contour. One chain end has $s = 0$ and the other has $s = N$, for a finite-sized chain of N monomers. The density is related to the position vector by

$$n(\vec{r}) = \int ds\ \delta(\vec{r} - \vec{R}(s))$$

This just counts the total number of monomers at real-space position $\vec{r}$. To calculate the correlation function, use the fact that for a Gaussian, random walk:

$$\left\langle \left(\vec{R}(s) - \vec{R}(s') \right)^2 \right\rangle = b^2 |s - s'|$$

where b is the step length of the walk. Also use the fact that for a one-dimensional Gaussian distribution:

$$\left\langle e^{iqx} \right\rangle = e^{-q^2 \left\langle |x|^2 \right\rangle /2}$$

5. Binodal and Spinodal

Derive a Landau expansion for the mean-field free energy of a binary mixture in terms of an expansion about the critical point, (ϕ_c, T_c) which is determined by the vanishing of both the second and third derivatives of the free energy with respect to ϕ. Calculate the binodal (coexistence curve in the ϕ, T plane) and the spinodal curve for this expansion.

Find the critical composition and temperature for a polymer solution and show how these quantities depend on the polymer molecular weight.

6. First-Order Transitions

When one studies phase transitions such as freezing, magnetism, and other order-disorder phenomena, the order parameter can be a nonconserved quantity, whose value in equilibrium is determined quite simply by the minimization of the Helmholtz free energy. Consider a Helmholtz free energy of the form:

$$f = -\frac{\epsilon}{2}\eta^2 + \frac{b}{3}\eta^3 + \frac{c}{4}\eta^4 + \dots \tag{1.159}$$

where η is the order parameter, b is nonzero and $c > 0$. Minimize f with respect to η and find the order parameter as a function of ϵ. Indicate the value of ϵ where the order parameter jumps from a value of $\eta = 0$ to a finite value. What is the jump in the order parameter?

7. Surfaces of Revolution

The location of a surface is defined by

$$\vec{r} = g(u)\cos v\,\hat{x} + g(u)\sin v\,\hat{y} + f(u)\hat{z} \tag{1.160}$$

Find the tangent vectors, normal, and area element. Show that if $g(u) = u$ and $f(u) = u\cot\alpha$ this represents a cone of angle α.

8. Relation Between Volume and Surface Area

Derive an expression for the volume of an object in terms of its surface area and normal, assuming the surface is closed. [*Hint*: Use the fact that $\nabla \cdot \vec{r} = 3$ and the divergence theorem.]

9. Curvatures in the Implicit Representation

Using the Monge form of the surface and the formulas for the mean and Gaussian curvatures for the implicit representation of a surface, *e.g.*, Eqs. (1.105,1.106), derive the expressions for H and K in terms of the derivatives of the height of the surface, $h(x, y)$.

10. Parallel Surfaces

Derive the relationships for the area elements and curvatures for parallel surfaces defined by translating the normal by a distance δ.

$$dA' = dA\left[1 + 2H\,\delta + K\,\delta^2\right] \tag{1.161}$$

$$H' = \frac{H + K\,\delta}{1 + 2H\,\delta + K\,\delta^2} \tag{1.162}$$

$$K' = \frac{K}{1 + 2H\,\delta + K\,\delta^2} \tag{1.163}$$

1.8 REFERENCES

1. L. D. Landau and E. M. Lifshitz, *Statistical Physics*, 3rd Edition, revised and enlarged by E. M. Lifshitz and L. P. Pitaevskii (Pergamon, New York, 1980).

2. S. K. Ma, *Statistical Mechanics* (World Scientific, Philadelphia, 1985).

3. M. Toda, R. Kubo, and N. Saito, *Statistical Physics I* (Springer-Verlag, New York, 1982).

4. P. M. Chaikin and T. C. Lubensky, *Principles of Condensed Matter Physics* (Cambridge University Press, Cambridge, 1995).

5. Vector methods are discussed by C. E. Weatherburn, *Differential Geometry of Three Dimensions* (Cambridge University Press, Cambridge, 1939).

6. L. D. Landau and E. M. Lifshitz, *Fluid Mechanics* (Pergamon, New York, 1982).

7. A. W. Adamson, *Physical Chemistry of Surfaces* (Wiley, New York, 1990).

8. C. A. Miller and P. Neogi, *Interfacial Phenomena* (Marcel Dekker, New York, 1985).

9. R. D. Vold and M. J. Vold, *Colloid and Interface Chemistry* (Addison-Wesley, Reading, MA, 1983).

10. W. B. Russel, D. A. Saville, and W. R. Schowalter, *Colloidal Dispersions* (Cambridge University Press, Cambridge, 1989).

11. P. G. de Gennes, *Scaling Concepts in Polymer Physics* (Cornell University Press, Ithaca, New York, 1979).

12. M. Doi and S. F. Edwards, *Theory of Polymer Dynamics* (Oxford University Press, Oxford, 1988).

13. For a survey of current research in surfactant science, refer to the series *Surfactants in Solution*, ed. K. Mittal (Plenum, New York); Volumes 1-11 have been published.

14. For a recent survey of the physics of amphiphilic systems, see *Physics of Amphiphilic Layers*, eds. J. Meunier, D. Langevin, and N. Boccara (Springer-Verlag, New York, 1987).

CHAPTER 2

Interfacial Tension

2.1 INTRODUCTION

One possible approach to the development of new materials focuses on the mixing of the components on the molecular level. The study of such mixtures involves the chemistry and the statistical mechanics related to the microscopic interactions. Another approach is to combine the materials on mesoscopic scales, tens to hundreds of Angstroms; for solids, such mixtures are termed **composites** and for liquids they are termed **dispersions**. Although the local environment of most of the molecules is that of the pure bulk components, the *macroscopic* properties of the system can nevertheless be drastically modified. A fundamental understanding of the processes involved in this second approach to materials design involves the physics of the interfaces between the two components. Sometimes, one does not want to make a bulk dispersion or composite material, but rather desires to modify the surface of a given material by adsorbing one or more layers of another component. An understanding of this process involves the physics of the surface — its structure and possible phase transitions — and the statistical mechanics of adsorption. We shall use the term **surface** to indicate the boundary between a fluid or solid in equilibrium with its vapor (or a vacuum if the vapor is dilute enough) and the term **interface** to denote the boundary between two different solids or liquids. It is important to note that an interface between two components is often *not* a sharp discontinuity between them; the interface usually has a width of at least a molecular thickness. In this chapter, we consider a model interface between two coexisting phases in equilibrium. We discuss the concept of interfacial tension and calculate the free energy and composition profile as one goes from one phase, through the interface, to the other phase. We also discuss how the tension may be modified by the presence of interfacially active components. For profiles that show a rather sharp transition between the two phases, interfaces and surfaces are quasi-two-dimensional and their physical properties often reflect the effects of this reduced dimensionality (*e.g.*, in the relatively large thermal fluctuations that roughen an interface or surface, as discussed in the following chapter).

2.2   FREE ENERGY OF SURFACES AND INTERFACES

Surface or Interfacial Tension

The additional free energy per unit area to remove molecules from the bulk and create interface between two coexisting phases is known as the **surface or interfacial tension**[1], γ, which has the units of energy per unit area (erg/cm^2), or force per unit length (dynes/cm). Both the energy and the entropy of the molecules are different at surfaces or interfaces and at the very least, the molecule loses entropy by being confined to the interface. A mechanical interpretation of the surface tension is that it is the two-dimensional analogue of pressure, force per unit length, parallel to the interface. Typical values of interfacial tensions are water-vapor $\gamma \approx 73$ dynes/cm, water-oil $\gamma \approx 57$ dynes/cm (depending on the type of oil), mercury-water $\gamma \approx 415$ dynes/cm.

Example: Dimensional Estimates of Interfacial Tensions

A rule of thumb in estimating interfacial tensions (based on the calculations to be described) is that the interfacial tension is a characteristic energy divided by a characteristic area related to size of the interfacial region. For simple fluids, this energy is often entropic in origin and is of order of magnitude the temperature T. When the interface has a microscopic size, say 3 Å, $\gamma \approx 40$ dynes/cm, comparable to typical water-oil tensions. On the other hand, for complex fluids, the relevant energy scales are still of order T (*e.g.*, related to the entropy loss of a particle confined to the interface, a value that is independent of its size), but the interface width may be much larger than a molecular size. For example, if the interface width is 100 Å, the interfacial tension is $\gamma \approx .04$ dynes/cm, a reduction by 3 orders of magnitude for a size increase of about 30. This sensitivity to the size comes from the quadratic dependence of γ on the interface width as shown in this chapter.∎

The interfacial tension of a stable, two-phase system is always positive, otherwise the two phases would spontaneously mix since they lower their free energy by making more and more interface. Therefore, near the critical point for phase separation, where the two coexisting phases become indistinguishable, one expects the surface tension between the two phases to vanish. The addition of a third interfacially active component to a two-component mixture with a tendency to phase separate can also result in an *effectively* negative tension (related to the chemical potential of the third component) which can cause the two components to spontaneously form a dispersion with an amount of internal interface related to the amount of the interfacially active component. Such systems are described in Chapter 8.

Laplace Pressure

Consider a curved interface with different pressures on each side (*e.g.,* a liquid in equilibrium with its vapor), as shown in Fig. 2.1. The free energy is

$$F = \gamma \int dA - p \int dV \tag{2.1}$$

where dA is the surface area element, dV is the volume element, and p is the pressure (Lagrange multiplier to conserve matter). For simplicity, take p in the gas phase to be negligible, since it is so dilute. Imagine displacing the surface by an amount δ parallel to the normal. Then using the result for parallel surfaces (the mean curvature, H, is taken to be negative for a sphere):

$$dA(\delta) = dA(0)(1 - 2\delta H + \delta^2 K) \tag{2.2}$$

and the relation that

$$dV(\delta) = dV(0) + \delta dA(0) \tag{2.3}$$

one can show that the change in the free energy to first order in δ is

$$F(\delta) - F(0) = \delta \int dA(0) \, [-2\gamma H - p] \tag{2.4}$$

In equilibrium, the free energy is stationary with respect to variations in the position of the interface; we thus require $dF/d\delta = 0$. This gives us the general relation for $\Delta p = p_1 - p_2$:

$$\Delta p = -2\gamma H \tag{2.5}$$

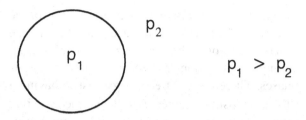

Figure 2.1 The Laplace pressure maintains a spherical interface in equilibrium.

relating the pressure difference across an interface to the curvature of the interface. Thus to maintain a curved interface, the pressure on the inside must exceed that of the outside. Of course, for a fixed volume (applicable to the incompressible, two-component systems that we consider), a sphere has the minimum surface area and hence the lowest surface energy and will be the equilibrium shape.

Surface Tension of Solids

For solids, the surface tension is *anisotropic* — it is different for different crystal faces, defined by their normal $\hat{n}$: $\gamma = \gamma(\hat{n})$. The equilibrium crystal shape is not a sphere, but is determined by the Wulff construction (for a proof see Ref. 2)

$$(\vec{r} \cdot \hat{n}) = \lambda\gamma(\hat{n}) \tag{2.6}$$

where λ is a constant analogous to the pressure (Lagrange multiplier to conserve volume). Given $\gamma(\hat{n})$ one obtains the equilibrium crystal shape, at zero temperature, by drawing the vectors $\vec{r}$ that satisfy this criterion, and $\vec{r} \cdot \hat{n}$ is the perpendicular distance from the origin to the tangent plane at the surface point $\vec{r}$. If the surface tension is isotropic, this is just the equation of a sphere.

2.3   SURFACE/INTERFACIAL TENSION THEORY

Interfacial Profile

The surface or interfacial tension is the free energy to make an interface in an otherwise uniform system. In a binary mixture of "A" and "B" molecules (or Ising model) one can imagine a system with a half-space of one phase (*e.g.*, "A" rich phase) in coexistence with a half-space of the other phase (*e.g.*, "B" rich). The equilibrium compositions are given by the coexistence curve (see Chapter

1) and these two compositions represent the minimum free energy states for the system with a fixed total composition. The transition in space from the "A" rich phase to the "B" rich phase is necessarily of higher energy since in going from "A" to "B" one has to pass through compositions for which the free energy is higher than that of the bulk phases. Therefore the free energy cost of having a smooth interfacial profile for the composition is related to the surface tension. In order to understand this effect quantitatively, even in a random-mixing or mean-field sense (*i.e.*, neglecting thermal fluctuations), one has to go beyond the simple treatment of Chapter 1 and consider the physics of the nonuniform system to derive the spatial profile and corresponding free energy. One of the difficulties is that within a simple lattice-gas model for the two components, the microscopic composition variable is discrete: $s_i = 0, 1$. Our intuition about the composition profile, however, requires a composition which varies in a continuous manner from the "A" rich phase (*e.g.*, small $\phi = \langle s_i \rangle$) to the "B" rich phase (*e.g.*, $\phi \sim 1$). We now present a variational method for formulating a mean-field theory that can take into account spatial dependence of the order parameter. This method, which is also used in later chapters, is then used to calculate the interfacial profile and tension.

Variational Method

Consider a system described by an exact Hamiltonian, $\mathcal{H}$. From Chapter 1, we know that the exact probability distribution, P, of the system is given by the Boltzmann factor, $P \sim e^{-\mathcal{H}/T}$; this arises from the minimization of the total free energy parameterized by the probability distribution, P:

$$\tilde{F} = T \int d\Lambda \, P \log P + \int d\Lambda \, P \, \mathcal{H} \tag{2.7}$$

where $\tilde{F}$ is the exact free energy and Λ represents the phase space of the system. A variational approximation to the true Boltzmann weight and hence to the free energy can be obtained by considering a model system, described by a Hamiltonian, $\mathcal{H}_0$, which contains several parameters and by minimizing Eq. (2.7) with respect to these parameters. Thus, the free energy is approximated by

$$F = T \int d\Lambda \, P_0 \log P_0 + \int d\Lambda \, P_0 \, \mathcal{H} \tag{2.8}$$

where $P_0 \sim e^{-\mathcal{H}_0/T}$. In fact, as shown below, the *exact* free energy, $\tilde{F}$, of the system with a Hamiltonian $\mathcal{H}$ obeys the inequality:

$$\tilde{F} < F = F_0 + \langle \mathcal{H} - \mathcal{H}_0 \rangle_0 \tag{2.9}$$

where F_0 is the free energy of the model system and the average values ($\langle ... \rangle_0$) are taken with respect to the Boltzmann factor $e^{-\mathcal{H}_0/T}$ of the model Hamiltonian. Thus, if one chooses $\mathcal{H}_0$ to have a given functional form with some unknown parameters, an approximation to the free energy may be derived by minimizing F as defined in Eq. (2.8) with respect to these parameters so the lowest upper bound[3] on $\tilde{F}$ is obtained.

Example: Proof of Variational Method

The proof of Eq. (2.9) begins by[3] considering two nonnegative distribution functions, $\tilde{P}(\vec{r})$ and $P_0(\vec{r})$, of the degrees of freedom, which we denote symbolically as an N-dimensional vector $\vec{r}$ ($\vec{r}$ is not necessarily the spatial coordinate). We also symbolically denote the sum over the degrees of freedom by the functional integral $\int d\vec{r}$. The distribution functions are normalized so that

$$\int d\vec{r}\, P_0(\vec{r}) = \int d\vec{r}\, \tilde{P}(\vec{r}) = 1 \qquad (2.10)$$

Now these two distribution functions satisfy the inequality

$$\int d\vec{r}\, P_0(\vec{r}) \log P_0(\vec{r}) \geq \int d\vec{r}\, P_0(\vec{r}) \log \tilde{P}(\vec{r}) \qquad (2.11)$$

This can be seen by showing that the difference

$$\int d\vec{r}\, P_0(\vec{r}) \log \left[\frac{P_0(\vec{r})}{\tilde{P}(\vec{r})} \right] \geq 0 \qquad (2.12)$$

since this equation can be rewritten as

$$\int d\vec{r}\, \tilde{P}(\vec{r}) \left[V \log V - V + 1 \right] \geq 0 \qquad (2.13)$$

where $V = P_0(\vec{r})/\tilde{P}(\vec{r})$. This follows because $\tilde{P} \geq 0$ by definition and because $x \log x \geq (x - 1)$ for any value of $x > 0$.

If we choose the normalized distribution functions:

$$P_0 = e^{\frac{(F_0 - \mathcal{H}_0)}{T}} \qquad (2.14a)$$

$$\tilde{P} = e^{\frac{(F - \mathcal{H})}{T}} \qquad (2.14b)$$

We see from Eq. (2.11) that Eq. (2.9) is satisfied where

$$F_0 = -T \log \int d\vec{r} \, e^{-\mathcal{H}_0/T} \tag{2.15}$$

$$\langle \mathcal{H} - \mathcal{H}_0 \rangle_0 = \int d\vec{r} \, (\mathcal{H} - \mathcal{H}_0) P_0 = \int d\vec{r} \, (\mathcal{H} - \mathcal{H}_0) \, e^{\frac{(F_0 - \mathcal{H}_0)}{T}} \tag{2.16}$$

■

Binary Mixture: Variational Approximation

In order to understand the composition profile and the interfacial free energy of a system with two coexisting phases, we must consider situations where the composition is not homogeneous — i.e., $\phi(\vec{r})$ varies in space. For example, in a system where the bulk equilibrium is between two phases with compositions ϕ_1, ϕ_2, one can investigate the situation where the composition asymptotically approaches ϕ_1 as $z \to -\infty$ and ϕ_2 as $z \to \infty$; the interface is located somewhere in the middle of these two regions. The simple, random mixing procedure described in Chapter 1 is no longer adequate to describe the spatial variation of the composition and we therefore use the variational procedure described above to derive an expression for the free energy of a binary mixture with a nonuniform composition. We begin with the lattice-gas model of a two-component system [see Chapter 1, Eq. (1.62)] and consider, for simplicity, only two-body interactions between the components. The Hamiltonian of the exact system, $\mathcal{H}$, is written as a function of the local composition variable at site i, s_i, where $s_i = 0$ represents the presence of the "A" species and $s_i = 1$ the "B" species. We then have

$$\mathcal{H} = \tfrac{1}{2} \sum_{ij} J_{ij} s_i (1 - s_j) \tag{2.17}$$

where J_{ij} is the *net* interaction between the two components (the signs are such that one gets phase separation when the $J_{ij} > 0$). The partition function for this Hamiltonian is difficult to evaluate because of the coupling between the sites. We take as the model Hamiltonian, $\mathcal{H}_0$, a function of single-site variables only:

$$\mathcal{H}_0 = \sum_i T \beta_i s_i \tag{2.18}$$

where the factor of T is put in the definition of β_i for convenience. The parameters β_i are determined using Eq. (2.9) to derive the upper bound, F, on the exact free energy, $\tilde{F}$, and then minimizing F with respect to the parameters β_i (with the constraint that the average composition is fixed) to determine the least upper bound and best estimate within the variational scheme. What was depicted before as an integral over all the degrees of freedom, $\int d\vec{p}d\vec{q}$, is here given by $\Pi_i \sum_{s_i=1,0}$ (*i.e.*, a product over all sites i of the sum over the two possible values of $s_i = 0, 1$). Note that because of boundary conditions on the compositions, the locally average compositions and hence the parameters β_i may vary in space. This is the case for a system with a surface or an interface.

The free energy of the model system, F_0 is given by

$$F_0 = -T \log Z_0 \qquad (2.19)$$

$$Z_0 = \Pi_i \sum_{s_i=1,0} e^{-\beta_i s_i} = \Pi_i \frac{1}{1 - \phi_i} \qquad (2.20)$$

where $\phi_i = (1 + e^{\beta_i})^{-1}$. One can show by taking the average with respect to $P_0 = e^{(F_0 - \mathcal{H}_0)/T}$ that $\phi_i = \langle s_i \rangle_0$; ϕ_i is the equilibrium average value of the local concentration variable, s_i in the ensemble described by $\mathcal{H}_0$ and P_0. Since it is an average quantity, ϕ_i can vary continuously from 0 to 1 and will be useful in constructing a continuous density profile for the system with an interface. A similar calculation yields

$$\langle \mathcal{H} - \mathcal{H}_0 \rangle_0 = \frac{1}{2} \sum_{ij} J_{ij}\phi_i(1 - \phi_j) - T \sum_i \beta_i \phi_i \qquad (2.21)$$

Using Eqs. (2.19,2.21) in Eq. (2.9) we find that the upper bound on the exact free energy is

$$F = \sum_i \left[T \left\{ (1 - \phi_i) \log(1 - \phi_i) + \phi_i \log \phi_i \right\} \right]$$

$$+ \frac{1}{2} \sum_{ij} J_{ij}\phi_i(1 - \phi_j) \qquad (2.22)$$

This has the same form as the free energy of a homogeneous system in the random mixing approximation; Eq. (2.22) generalizes this result to a system where the composition is *not* homogeneous and the local value of ϕ_i changes in space. Regarding now the ϕ_i as the variational parameters, one minimizes F with respect to ϕ_i, respecting the boundary conditions and maintaining the constraint of fixed total composition (*i.e.*, in practice, one minimizes $G = F - \sum_i \mu\phi_i$). For a system with a surface or an interface, the boundary conditions will imply that all of the ϕ_i are *not* identical and the minimization will determine the **composition**

profile. For practical problems, this is most effectively done in the continuum limit since the finite difference equations obtained from the minimization of Eq. (2.22) can be written in terms of differential equations.

Continuum Limit

The continuum limit of Eq. (2.22) is obtained by noting that

$$J_{ij}\phi_i(1 - \phi_j) = \tfrac{1}{2} J_{ij} \left[(\phi_i - \phi_j)^2 - \phi_i^2 - \phi_j^2 + 2\phi_i \right] \qquad (2.23)$$

Going over to a continuum notation for a free energy per unit volume (instead of a free energy per site) allows us to convert the difference $\phi_i - \phi_j$ to a gradient. The precise form of this gradient depends on the coupling matrix J_{ij}. For the case of short-range, nearest-neighbor interactions, we can write

$$(\phi_i - \phi_j) \rightarrow a\nabla\phi \qquad (2.24)$$

where a is the distance between nearest neighbors. We then have the continuum version of Eq. (2.22) with $\phi_i \rightarrow \phi(\vec{r})$ being the *local* value of the composition and the total free energy is given by

$$F = \int d\vec{r} \left[f_0[\phi(\vec{r})] + \tfrac{1}{2} B |\nabla\phi(\vec{r})|^2 \right] \qquad (2.25)$$

where $B = J/(2a)$ where $J = \sum_j J_{ij}$. The nonlinear, local part of the free energy per unit volume, f_0 is

$$f_0[\phi] = \frac{1}{a^3} \left(T \left[\phi \log \phi + (1 - \phi) \log(1 - \phi) \right] + \frac{J}{2} \phi(1 - \phi) \right) \qquad (2.26)$$

For a uniform system, this is the only contribution to the free energy per unit volume; for the nonuniform system one evaluates f_0 with the *local* value of $\phi(\vec{r})$.

Thus, our variational treatment in the continuum limit gives the free energy as a functional (involving both functions and gradients) of the local, average composition, $\phi(\vec{r})$. To proceed to find the spatial variation of $\phi(\vec{r})$ one must minimize F functionally with respect to $\phi(\vec{r})$ with the appropriate boundary conditions. For the form of f_0 written earlier there is no simple analytic solution to these equations. However, in the limit that the composition is close to the critical composition, $\phi = 1/2$, one can expand Eq. (2.26) about $\psi = \phi - 1/2$ to obtain

$$f_0 \approx \frac{1}{a^3} \left[T \left(2\psi^2 + \frac{4\psi^4}{3} - \log 2 \right) + \frac{J}{2}(1/4 - \psi^2) \right] \qquad (2.27)$$

The gradient term has the same form as before, but with $\nabla \psi$ instead of $\nabla \phi$ since ϕ and ψ are just related by a constant. With the approximation in Eq. (2.27) we can write:

$$F = \int d\vec{r} \left[-\frac{\epsilon}{2}\psi(\vec{r})^2 + \frac{1}{4}c\psi(\vec{r})^4 + \frac{B}{2}|\nabla\psi(\vec{r})|^2 \right] \qquad (2.28)$$

where constant terms have been dropped and linear terms in $\phi(\vec{r})$ are omitted since they are incorporated in the chemical potential that constrains the average composition (*i.e.*, $\int d\vec{r}\,\phi(\vec{r})$ is fixed). The coefficients are related to the microscopic parameters by $B = J/(2a)$, $\epsilon = (J - 4T)/a^3$, and $c = 16T/(3a^3)$. In particular, ϵ vanishes at the critical point given by $T_c = J/4$; this is the same value found in the discussion in Chapter 1. This expansion of the free energy, including the gradient terms, is known as a **Ginzburg-Landau expansion**.

Free Energy Minimization

The Helmholtz free energy, F, of Eq. (2.28) must be minimized to find the composition for a system with an interface — *e.g.*, we look for solutions where the composition at $z \to \infty$ is equal to one of the bulk, equilibrium values (*e.g.*, ϕ_1 in the notation of Chapter 1) and the composition at $z \to -\infty$ is equal to the value of the coexisting phase with composition ϕ_2. However, this minimization must be performed with the constraint that the composition, averaged over the *two* coexisting phases is fixed. This is done by considering the grand potential, G, as described in Chapter 1. For the case where the average composition is equal to the critical composition, the additional term due to the chemical potential vanishes and $F = G$. We therefore consider this simple case and look for solutions that functionally minimize the Helmholtz free energy, F. One can explicitly confirm that the solutions we derive below conserve the average composition to equal its critical value — *i.e.*, $\int d\vec{r}\,\psi(\vec{r}) = 0$.

To minimize the approximate free energy of Eq. (2.28), we use the result from the calculus of variations[4] that minimizes a functional, F, which is an integral over a spatially varying function, $\psi(\vec{r})$, with respect to all possible variations of this functions. We write:

$$F = \int d\vec{r}\, f\left[\psi(\vec{r}), \nabla\psi\right] \qquad (2.29)$$

The function, $\psi(\vec{r})$, which minimizes F is determined by the equation:

$$\frac{\delta F}{\delta \psi(\vec{r})} = \frac{\partial f}{\partial \psi} - \frac{\partial}{\partial r_i}\frac{\partial f}{\partial \psi_{r_i}} = 0 \qquad (2.30)$$

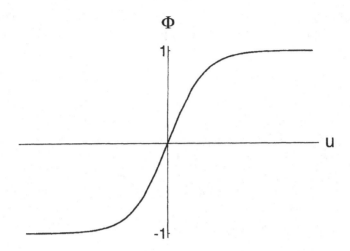

Figure 2.2 Interfacial profile where Φ is equal to the order parameter, ψ, scaled to its bulk value, ψ_0, and $u = z/\xi$, the bulk correlation length. The width of the interfacial region centered at $u = 0$ scales with ξ.

where $\psi_{r_i} = \partial\psi/\partial r_i$. In Eq. (2.30) one sums over the repeated index i and $\vec{r}_i = (x, y, z)$. The resulting equation is

$$-\epsilon\psi + c\psi^3 - B\nabla^2\psi = 0 \tag{2.31}$$

with the boundary conditions that the system is uniform away from the interface. Far away from the interface, we expect the two phases to have their equilibrium values of the composition. In the present approximation that $\psi = \phi - 1/2$ is small, these values are given by

$$\psi_0 = \pm\sqrt{\frac{\epsilon}{c}} \tag{2.32}$$

Interfacial Profile

Consider a one-dimensional concentration variation, $\psi(z)$, with boundary conditions $d\psi/dz = 0$ at $z = \pm\infty$, which ensure the equilibrium phases at these limits. The solution is then

$$\psi(z) = \sqrt{\frac{\epsilon}{c}} \tanh\frac{z}{\xi} \tag{2.33}$$

where the width of the interfacial region is given by $\xi = \sqrt{2B/\epsilon}$, which is the bulk correlation length and diverges as $\epsilon \to 0$ or $T \to T_c$. Thus, for $z \to \pm\infty$, ϕ approaches its equilibrium, uniform values of $\pm\sqrt{\epsilon/c}$ as shown in Fig. 2.2. The width of the interfacial region is proportional to $1/\sqrt{\epsilon}$, so that as the critical point is approached ($\epsilon \to 0$) the interfacial width diverges. To find the surface tension or interfacial tension, one inserts this solution back into the free energy per unit area and subtracts off the free energy per unit area of the bulk phases (where ϕ is a constant). Thus:

$$\gamma = \int_{-\infty}^{\infty} dz \left(f_0[\psi(z)] - f_0[\psi_0] + \tfrac{1}{2}B\psi_z^2 \right) \tag{2.34}$$

where f_0 is the local free energy per unit volume defined in Eq. (2.27), $\psi_z = \partial\psi/\partial z$, and ψ_0 is the bulk, equilibrium order parameter of Eq. (2.32). The free energy of the inhomogeneous system with an interface differs from the free energy of the homogeneous system by (i) the fact that the bulk free energy terms are not equal and by (ii) the presence of a gradient energy term. To simplify the calculation, note that an integration of Eq. (2.31) in one dimension, with the boundary conditions $d\psi/dz = 0$ at both plus and minus infinity where $\psi = \psi_0$, implies that

$$f_0[\psi(z)] - f_0[\psi_0] = \tfrac{1}{2}B\psi_z^2 \tag{2.35}$$

Thus Eq. (2.34) becomes

$$\gamma = B \int_{-\infty}^{\infty} dz \, \psi_z^2 \tag{2.36}$$

Thus, the free energy cost of the interface can be expressed in terms of the gradient energy term (which of course balances the change in the bulk free energy due the presence of the interface). For the profile given by Eq. (2.33), this yields an interfacial tension

$$\gamma = \frac{J(T_c - T)}{2Ta\xi} \tag{2.37}$$

which vanishes as $T \to T_c$ since the two coexisting phases become identical. Note that in this mean-field theory, the surface tension vanishes as $\sqrt{(T_c - T)^3}$ since the correlation length, $\xi \sim 1/\sqrt{T_c - T}$.

Example: Gas-Liquid Interface

Although Eq. (2.28) has been derived in the context of the lattice-gas (Ising) model for a binary system with a critical composition of 1/2, it can be reinterpreted quite generally for a liquid-gas system. Derive this form from the virial expansion for an imperfect gas and show that the difference in the grand potential per unit volume, $g(n)$, between a system with a local density of n and the grand potential of the bulk gas or liquid system, g_b, at coexistence (where the liquid and gas have equal grand potentials), has the form:

$$g(n) - g_b = g_0(n - n_\ell)^2(n - n_v)^2 \tag{2.38}$$

where n_ℓ and n_v are the densities of the liquid and gas (vapor) at equilibrium and g_0 is a constant.

We consider a virial expansion for the Helmholtz free energy per unit volume, f, of an imperfect gas as discussed in Chapter 1:

$$f(n) = T \left[n \left(\log nv_0 - 1 \right) + \tfrac{1}{2}an^2 + \tfrac{1}{3}bn^3 \right] \tag{2.39}$$

where a has the dimensions of a volume and b has dimensions of volume squared. As discussed in Chapter 1, the critical value of the density and the temperature (corresponding here to the critical value of the virial coefficient, a) is defined by the density at which $\partial^2 f/\partial n^2 = \partial^3 f/\partial n^3 = 0$. Thus, we find $n_c = 1/\sqrt{2b}$ and $a_c = -2\sqrt{2b}$. We now expand the free energy of Eq. (2.39) about $n = n_c$ with $a \approx a_c$. We find

$$f(n) - f(n_c) \approx \left[h(n - n_c) - \tfrac{1}{2}\epsilon(n - n_c)^2 + \tfrac{1}{4}c_0(n - n_c)^4 \right] \tag{2.40}$$

where $h = T(\log n_c v_0 + an_c + bn_c^2)$, $\epsilon = -T(a - a_c)$, and $c_0 = T(3n_c^3)^{-1}$, with dimensions of the product of an energy and a volume cubed.

To find the coexisting densities we recall the conditions for coexistence from Chapter 1: the equality of the chemical potentials and osmotic pressures. Consider a system where the average density (*i.e.*, averaged over the two coexisting phases) is equal to n_c; this corresponds to fixing the chemical potential to a value $\mu = \left[\partial f/\partial n\right]_{n_c}$. For this case, $\mu = h$ and the conditions for coexistence imply that the two coexisting phases are given by the minimization of the grand potential: $\partial g/\partial n = 0$ where

$$g(n) = f(n) - f(n_c) - \mu(n - n_c)$$

$$= \left[-\tfrac{1}{2}\epsilon(n - n_c)^2 + \tfrac{1}{4}c_0(n - n_c)^4 \right] \tag{2.41}$$

We find one root — the liquid — with a density, $n_\ell = n_c + \sqrt{\epsilon/c_0}$ and the other root at lower density — the gas — with a density $n_v = n_c - \sqrt{\epsilon/c_0}$. It is easily seen that $g(n_v) = g(n_\ell) = g_b$ and if we define $W = [g(n) - g_b]$, we find that we can write, with $c = c_0/4$:

$$W(n) = (n - n_\ell)^2 (n - n_v)^2 \qquad (2.42)$$

In finding the shape of the profile between the two phases it is $W(n)$ that is used as the local free energy since the minimization occurs at constant chemical potential (equivalent to constant *average* density), hence the subtraction of the term μn in g and W. Furthermore, the interfacial free energy is the difference between the free energy of the system with an interface and that of the bulk fluid or gas, hence the subtraction of g_b to arrive at $W(n)$. Either the analogy to the lattice-gas model or the density functional expansion discussed briefly in Chapter 1, suggests that for systems with short-range interactions, the gas-liquid profile is determined by the minimization of the interfacial free energy:

$$G_s = \int d\vec{r}\, \left[W[n(\vec{r})] + \tfrac{1}{2}B|\nabla n(\vec{r})|^2 \right] \qquad (2.43)$$

where B has the dimensions of an energy multiplied by a length to the fifth power. In a more microscopic theory, such as that suggested by the density functional expansion, this length would be related to the range of the attractive interactions. Considering a one-dimensional density variation from the gas $(n(z \to -\infty) \to n_v)$ to the liquid $(n(z \to \infty) \to n_\ell)$ and minimizing as before, we find

$$\tfrac{1}{2}B \left(\frac{\partial n}{\partial z} \right)^2 = c(n - n_v)^2(n - n_\ell)^2 \qquad (2.44)$$

This equation satisfies the boundary conditions that as $z \to \pm\infty$ the system takes on the equilibrium values of the order parameter. Taking the square root (determining the signs by noting that $n_\ell > n > n_v$) and integrating, we find that the profile centered at $z = 0$ is given by

$$\log \left[\frac{n_\ell - n}{n - n_v} \right] = -\frac{z}{\xi} \qquad (2.45)$$

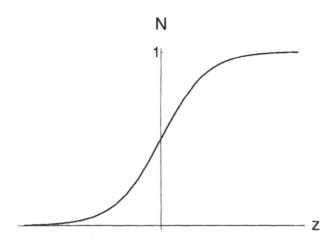

Figure 2.3 Interfacial profile for gas-liquid system where $N = n/n_\ell$ (the density scaled to the equilibrium liquid density) and $Z = z/\xi$ as defined in the text. The curve is calculated for the case where $n_v \ll n_\ell$.

where $\xi^{-1} = (n_\ell - n_v)\sqrt{2c/B}$ has the dimensions of an inverse length. This expression can be solved to give

$$n(z) = \frac{n_\ell}{1 + e^{-z/\xi}} + \frac{n_v}{1 + e^{z/\xi}} \tag{2.46}$$

which describes the interface between the gas at negative values of z and the liquid at positive values of z. A sketch of the profile is shown in Fig. 2.3. Near the critical point, where $n_\ell - n_v \to 0$, the width of the interfacial region, characterized by ξ, diverges. Note that this simple situation is valid for a system whose chemical potential is equal to the critical chemical potential so there are no linear terms in g. ∎

2.4   SURFACE-ACTIVE AGENTS

Surfactants

The interfacial tension between two species can be reduced by the addition of third component whose free energy is lowered when it resides on the surface. Examples of such agents are surfactants that consist of molecules with both polar and nonpolar groups joined together chemically. Anionic surfactants have a counterion that is soluble in a polar liquid (such as water) and is positively charged (*e.g.*, $R_n OSO_3 Na^+ SO_3^-$ where the Na^+ is water soluble and R_n signifies the hydrocarbon chain). Cationic surfactants have negative counterions (*e.g.*, Br_2^-) and nonionic surfactants have a polar group, but no ion in solution (they tend to have a polar group like OCH_2CH_2OH).

Reduction of Surface Tension

Consider a surfactant that is not soluble in the solution and that behaves like an ideal gas on the surface of a fluid with area A. The total surface energy is

$$F_s = A\left(\gamma_0 + T\sigma(\log \sigma a_0 - 1) + u_0\sigma\right) \qquad (2.47)$$

The first term is the bare surface energy of the fluid, the second term, the ideal gas entropy of the surfactant whose area density is, σ (a_0 is a constant with dimensions of a molecular area, comparable to the quantity v_0 for a dilute, three-dimensional gas), and the third term is the difference in energy of the surfactant molecule on the surface compared, for example to its energy in a solid phase of pure surfactant. (We assume here that there are no molecules in the bulk solvent.) In the previous section, Eq. (2.34), we considered the case of an equilibrium interface with no additional surface-active components. The surface tension, γ is defined by the difference in free energy between a system with two coexisting phases (and hence an interface) compared with the bulk free energies. When one has surface-active components, one must add to this surface energy additional terms that represent the free energy of these components at the interface. The surface tension in the system of fluid + surfactant is given by the change in the free energy as the amount of interfacial area is varied:

$$\gamma = \frac{\partial F_s}{\partial A} \qquad (2.48)$$

Now, $\sigma = N_s/A$, where N_s is the number of surfactant molecules on the surface, which is fixed. Taking the derivative including the variation of σ, we find that

$$\gamma = \gamma_0 - T\sigma \qquad (2.49)$$

This reduction in the tension is due to the fact that as the interfacial area is increased, the translational entropy of the surface active component is increased (for a fixed number of the surface active molecules). This increase in entropy lowers the system free energy and thus reduces γ from its bare value. This expression is correct for small area fractions of the surface active species, since we considered only the ideal-gas entropy.

Example: Excluded Volume Effects on Surface Tension

Find an expression for the surface tension of a fluid with an adsorbed surfactant layer with arbitrary area fraction, ϕ, for the case of insoluble surfactants.

The excluded volume interactions between the surfactants will *enhance* the lowering of the interfacial tension, compared with the ideal-gas case discussed previously. For example, one can consider a lattice-gas model for the adsorbed layer. In a mean-field approximation, the total free energy, F_s, is written:

$$F_s = N \left[T \left(\phi \log \phi + (1 - \phi) \log(1 - \phi) \right) \right] \qquad (2.50)$$

Here, $N = A/a^2$, is the total number of surface sites, where A is the total area and a^2 is the area per surfactant; γ_0 is defined above. Note that since the area fraction, $\phi = \sigma a^2$, (where σ is the area density), for a *fixed* number of surfactant molecules depends on A, one must take derivatives of ϕ into account when computing γ from Eq. (2.48). The result is

$$\gamma = \gamma_0 + \frac{T}{a^2} \log(1 - \phi) \qquad (2.51)$$

In the limit of small surface density, $\phi = \sigma a^2 \ll 1$, this expression reduces to Eq. (2.49) derived in the ideal-gas approximation. This reduction is larger than that predicted from Eq. (2.49); the change in entropy due to excluded volume interactions between the surfactant molecules enhances their effectiveness in lowering the surface tension. One can include other types of interactions between surfactant molecules (*e.g.*, attractions) in a similar manner.   ∎

Soluble Surfactants

The ability of the surfactant to increase its translational entropy lowers the surface tension. However, we note that this expression is only valid for *insoluble* surfactants at low concentrations, so the tendency for γ to become negative at large σ is just an indication that these approximations are breaking down. For soluble surfactants, one cannot consider fixed N_s. Rather, one has to equate the chemical potentials of the surfactants on the surface and in the bulk. For small concentrations, the surface tension is still reduced in a linear manner, but at large concentrations, the reduction in surface tension saturates due to the formation of micelles in the bulk; this is discussed in Chapter 8.

2.5 PROBLEMS

1. Composition Fluctuations

To calculate the correlation functions for composition fluctuations in a binary system, consider the "Ising" Hamiltonian of Eq. (2.17) with an additional term $\sum_i h_i s_i$ where h_i is a local field. Show that the correlation function

$$\langle (s_i - \langle s_i \rangle)(s_j - \langle s_j \rangle) \rangle = -T \frac{\partial^2 F[\{h_i\}]}{\partial h_i \partial h_j}$$

where the derivatives are evaluated at $h_i = 0$ for all sites i and the $F[\{h_i\}]$ is given by the expression for F in Eq. (2.22) supplemented by a term $\sum_i h_i m_i$. Using this expression for F calculate the correlation function in Fourier space: $G(\vec{q}) = \langle s(\vec{q}) s(-\vec{q}) \rangle$ as a function of the Fourier transform of the interaction matrix,

$$J(\vec{q}) = \sum_j J_{ij} e^{i\vec{q} \cdot \vec{R}_{ij}}$$

How does $G(\vec{q})$ behave for temperatures and compositions near the critical point as defined in Chapter 1?

Use the continuum description of the binary system described by the free energy of Eq. (2.28) to derive an expression for the correlation function for composition fluctuations. To do this, treat Eq. (2.28) as a new effective Hamiltonian. Examine the properties of the correlation function near the critical point and compare with the results of the "lattice-gas" model of Eq. (2.22). Which fluctuations does the continuum model best describe?

2. Interfacial Profile of Polymer Interfaces

Consider a polymer system that shows phase separation into a phase with a high density of polymer in solution that coexists with a phase with a low density of polymer in solution. For this system, one can write the Helmholtz free energy of the form:

$$F = \int d\vec{r} f \tag{2.52}$$

$$f = -\frac{\epsilon}{2}\phi(\vec{r})^2 + \frac{c}{3}\phi(\vec{r})^3 + \frac{A}{2\phi(\vec{r})}|\nabla\phi(\vec{r})|^2 \tag{2.53}$$

where ϕ is the polymer concentration. The first term comes from the net attraction of the polymer segments to each other, the second comes from the excluded volume repulsion of the segments; these are the first two terms in a virial expansion in the density where the translational entropy of the polymer $f_t \sim (\phi/N)\left[\log[\phi/N] - 1\right]$ is neglected for large molecular weights, N. The gradient term comes from the segment interactions and the factor of $1/\phi$ in the coefficient comes from the chain connectivity requirement[5]. When one examines the bulk coexistence curve for phase separation, including the contribution from f_t, one finds that the dilute phase has a volume fraction that is very small and vanishes as $N \rightarrow \infty$.

First, consider the bulk system where the gradient term can be neglected. The conditions for equilibrium of a dilute phase (where $\phi \approx 0$) and a dense phase (with a finite value of ϕ) are (i) equal chemical potentials $(\partial f/\partial \phi)$ and (ii) equal osmotic pressure

$$\Pi = \frac{\phi^2}{v_0}\left(\frac{\partial(f/\phi)}{\partial\phi}\right) \tag{2.54}$$

where v_0 is a molecular volume. If the dilute phase is approximated as $\phi = \Pi = 0$ find the value of ϕ in the coexisting phase from Eq. (2.54) and an approximate value of μ for this equilibrium. These approximations are correct in the limit of large molecular weights where the translational entropy of the dilute phase can be neglected.

Using this value for μ, minimize the free energy of Eq. (2.53) with the constraint of conservation by adding to Eq. (2.53) a term $-\mu\phi$ (i.e., minimize the grand potential). Next, include the gradient term for the case of a one-dimensional concentration variation and find the spatial dependence of the profile and the interfacial tension. [Hint: To reduce the complexity of the gradient term make the transformation: $\psi = \sqrt{\phi(\vec{r})}$ and minimize with respect to ψ.]

3. Surface Tension for Soluble Surfactants

Derive an expression for the surface tension for a system where the molecules adsorbed in a single monolayer on the surface are described by a lattice gas with the free energy of Eq. (2.50), but where the molecules in the bulk solvent interact with either attractive or repulsive forces described by a dilute, nonideal gas (see Chapter 1). From the requirement that these two systems (surface and bulk) must be in equilibrium and thus have the same chemical potential, derive an expression for the volume fraction adsorbed on the surface as a function of the volume fraction of molecules in the bulk.

2.6 REFERENCES

1. J. S. Rowlinson and B. Widom, *Molecular Theory of Capillarity* (Clarendon Press, Oxford, 1982).

2. G. C. Benson and D. Patterson, *J. Chem. Phys.* **23**, 670 (1955).

3. J. P. Hansen and I. R. McDonald, *Theory of Simple Liquids* (Academic Press, New York, 1990), p. 152; A. Isihara, *J. Phys. A* **1**, 539 (1968).

4. G. Arfken, *Mathematical Methods for Physicists* (Academic Press, NY, 1985).

5. P. G. de Gennes, *Scaling Concepts in Polymer Physics* (Cornell University Press, Ithaca, New York, 1979), p. 254.

CHAPTER 3

Fluctuations of Interfaces

3.1 INTRODUCTION

In thermal equilibrium, the surfaces of fluids and even solids are not perfectly flat; neither are the interfaces between two different phases. Thermal fluctuations roughen the regions between two materials; how strong these effects are depends on dimensionality (*i.e.,* if the interface is a line or a plane) and on temperature. For fluid interfaces, two-dimensional systems with one-dimensional interfaces have a mean-square roughness of the interface that varies linearly with the system size; the corresponding quantity in three-dimensional systems is logarithmic with the system size. For solids, the periodic potential tends to decrease the roughness of the interface; however, above the roughening transition temperature, which is typically some fraction of the melting temperature, the fluctuations of the position of the interface resemble those of a liquid. In this chapter, we first derive the free energy of an interface that is not necessarily flat, from the free energy of the two phases bounding this interface. We calculate the magnitude of the thermal fluctuations and discuss the roughening transition of solids. In the examples, we discuss two examples of unstable interfaces; in one (Rayleigh instability of a cylinder) the surface tension drives the instability and in the other (rupture of a thin film), the surface tension opposes the instability.

3.2   FREE ENERGY OF A FLUCTUATING INTERFACE

Interface Free Energy

In order to study the fluctuations of interfaces and surfaces, one first needs an expression for the free energy of an interface with a shape that is not necessarily flat. Phenomenologically, one can write that the free energy is the product of the interfacial tension, γ, and the area, which in the Monge representation for the surface, $z = h(x, y)$, reads:

$$F_s = \gamma \int dx dy \sqrt{1 + h_x^2 + h_y^2} \qquad (3.1)$$

where $h_x = \partial h / \partial x$. As we shall show below this is correct for interfaces where the spatial variation of the interface position occurs over length scales much larger than the typical width of the interface. It is only in this limit — frequently, the important limit in many physical situations where the interface is well defined to within a molecular scale — that one can study the long-wavelength undulations of the interface using the simple surface tension expression of Eq. (3.1).

Derivation of Interfacial Free Energy

A more rigorous derivation of this result[1], starting from the free energy as a function of the composition in a system with two coexisting phases, follows the discussion in Ref. 1. The free energy due to different configurations of the interface between these phases is calculated in the context of the three-dimensional binary system. The model for the free energy of the two coexisting bulk phases is the continuum, Ginzburg-Landau expression discussed in Chapter 2; this description originates from an Ising model representation of a binary mixture. We first consider Eq. (2.25) for the free energy ($\int d\vec{r} = \int dx dy dz$) as a function of the composition, $\phi(\vec{r})$:

$$F = \int d\vec{r} \, \left[\tfrac{1}{2} B (\nabla \phi)^2 + f_0[\phi(\vec{r})] \right] \qquad (3.2)$$

where B is a constant. As discussed in Chapter 2, the first term arises from the spatial gradients of the order parameter whose microscopic origin are the interactions in the system, while the second is the local free energy per unit volume of the bulk. The composition profile is taken to have the form

$$\phi(\vec{r}) = \chi \, (x, y, z - h(x, y)) \qquad (3.3)$$

where we show below that χ is related to the solution of the one-dimensional interface problem, $\phi = M(z)$, which obeys

$$\frac{\partial f_0}{\partial M} = B M_{zz} \tag{3.4}$$

where $M_z = \partial M/\partial z$. For the fluctuating interface, Eq. (3.3) indicates that we expect the solution to resemble that of the one-dimensional problem, but with an interface located at $z = h(x, y)$. We shall find a variational approximation to the solution by considering Eq. (3.3) as the solution of the Euler-Lagrange equation for the functional minimization of the free energy of Eq. (3.2).

Minimization of the Free Energy

To solve for the profile that minimizes F, one must solve the Euler-Lagrange equation taking into account the variations in $\chi (x, y, z - h(x, y))$ in three dimensions:

$$\left[1 + h_x^2 + h_y^2\right] \chi_{zz} + \left[\chi_{xx} + \chi_{yy}\right] - \left[h_{xx} + h_{yy}\right] \chi_z$$

$$- \frac{1}{B} \frac{\partial f_0}{\partial \chi} - 2 \left[h_x \chi_{zx} + h_y \chi_{yz}\right] = 0 \tag{3.5}$$

In our approximate analysis, we assume that the spatial variations in the $\hat{z}$ direction are much larger than those in the $\hat{x}$ or $\hat{y}$ directions. We thus keep only those terms proportional to the second derivative in z. This is equivalent to assuming that the spatial variations of the interface position occur over length scales much longer than the interface thickness. One then has the approximate equation for χ,

$$B \left[1 + h_x^2 + h_y^2\right] \chi_{zz} = \frac{\partial f_0}{\partial \chi} \tag{3.6}$$

But this is just the equation obeyed by the one-dimensional solution $M(z)$ (see Eq. (3.4)) if we replace z by $(z - h)/[1 + h_x^2 + h_y^2]^{1/2}$, so we find that

$$\chi \approx M \left(\frac{z - h(x, y)}{\sqrt{1 + h_x^2 + h_y^2}} \right) \tag{3.7}$$

The physical meaning of this expression is that to a first approximation, valid when the interface varies slowly in the $\hat{x}$ and $\hat{y}$ directions compared to its width in the $\hat{z}$ direction, is that the composition is just $M(z)$ calculated with an argument that represents the *normal* component of the distance to the interface:

$$M\left((\vec{r} - \vec{R}) \cdot \hat{n}\right) \tag{3.8}$$

where $\vec{r} = (x, y, z)$ is the spatial variable, $\vec{R} = (x, y, h(x, y))$ defines the interface and

$$\hat{n} = \frac{\hat{z} - h_x \hat{x} - h_y \hat{y}}{\sqrt{1 + h_x^2 + h_y^2}} \tag{3.9}$$

is the normal to the interface. Multiplying Eq. (3.6) by χ_z and using the boundary condition that the profile is flat far away from the interface where the free energy is that of the bulk — *i.e.*, $\chi_z = 0$ when the composition attains its equilibrium bulk value, ϕ_b, which is given by the minima of the bulk free energy, $f_0(\phi)$. Writing $f_b = f_0(\phi_b)$ as the free energy of the bulk, one finds

$$B\left[1 + h_x^2 + h_y^2\right]\chi_z^2 = 2(f_0(\chi) - f_0(\phi_b)) \tag{3.10}$$

Using this equation, one can find the surface free energy, F_s, by subtracting from the free energy of the system with a single interface the free energy of the bulk system with no interface — *i.e.*, $F_s = F - F_b$ where F_b is the total bulk free energy ($F_b = \int d\vec{r}\, f_b$). We find an expression identical to Eq. (3.1), where now the surface tension γ is determined from the one-dimensional profile:

$$\gamma = B\int_{-\infty}^{\infty} dz\, [M_z(z)]^2 \tag{3.11}$$

This is exactly the same expression for the energy of the flat interface determined in Chapter 2. The factor under the square root sign in Eq. (3.1) comes from the area of the plane perpendicular to the normal given by Eq. (3.9). We have thus shown that the energy of the interface is essentially given by the product of the surface tension (determined consistently from the composition profile for the flat interface) and the area of the undulated interface. This expression is valid in the approximation that the interface width (*e.g.*, variation of the profile in the $\hat{z}$ direction) is much narrower than the spatial variations of the interface in the perpendicular (*e.g.*, in the $\hat{x}$ and $\hat{y}$ directions).

3.3 THERMAL FLUCTUATIONS OF INTERFACES

Undulated Interfaces

We now consider the properties of fluctuating interfaces. The simplest case to consider is where the fluctuations are thermal in origin; the mean-square amplitude is then proportional to the temperature relative to the surface tension energy. These spontaneous undulations of interfaces can be observed by measurements of the interface thickness which are sensitive to the height-height correlations of the interface. For fluid interfaces, in the limit of negligible viscosity, these fluctuations are known as capillary waves and their dynamical properties are studied by light-scattering experiments[2,3]. In this section, we calculate the mean-square amplitude of these fluctuations and the mode frequencies corresponding to capillary waves on the surface of a fluid in the limit of negligible viscosity.

Free Energy of Fluctuations

Consider the fluctuations of a surface defined in the Monge representation as $z = h(x, y)$. The area of the flat surface is denoted by A. For slowly varying fluctuations of this surface about a flat shape ($h = h_0$, where h_0 is a constant) the additional surface free energy of the undulated interface over that of the flat one ($\Delta F_s = F_s - \gamma A$) is approximately

$$\Delta F_s = \tfrac{1}{2}\gamma \int dxdy \left(h_x^2 + h_y^2 \right) \tag{3.12}$$

In Fourier coordinates we define

$$h(\vec{\rho}) = \frac{1}{\sqrt{A}} \sum_{\vec{q}} h(\vec{q})\, e^{i\vec{q}\cdot\vec{\rho}} \tag{3.13}$$

$$h(\vec{q}) = \frac{1}{\sqrt{A}} \int d\vec{\rho}\, h(\vec{\rho})\, e^{-i\vec{q}\cdot\vec{\rho}} \tag{3.14}$$

where both $\vec{\rho}$ and $\vec{q}$ are two-dimensional. We then have the expression for the surface energy:

$$\Delta F_s = \tfrac{1}{2}\gamma \sum_{\vec{q}} q^2 \, |h(\vec{q})|^2 \tag{3.15}$$

Treating ΔF_s as a Hamiltonian of the fluctuating variable $h(\vec{q})$, the probability to find $h(\vec{q})$ with any particular value is proportional to $\exp[-\Delta F_s/T]$, which in our approximation is a Gaussian where all the different $\vec{q}$ modes are independent.

Using the relationship between the probability distribution and the mean-square value of a fluctuating quantity (see Chapter 1), we have

$$\langle |h(\vec{q})|^2 \rangle = \frac{T}{\gamma q^2} \tag{3.16}$$

This is the mean-square value of each Fourier mode in thermal equilibrium.

Real-Space Fluctuations

The mean-squared, real-space fluctuations of the surface about a flat profile, are given by

$$\langle h^2(\vec{\rho}) \rangle = \frac{1}{A} \sum_{\vec{q}} \langle |h(\vec{q})|^2 \rangle = \frac{1}{(2\pi)^2} \int d\vec{q} \, \langle |h(\vec{q})|^2 \rangle \tag{3.17}$$

Using Eq. (3.16) and performing the integral, we see that one must be careful about the lower limit of the integral because of the logarithmic divergence (when $\vec{q}$ is two-dimensional). Thus introducing a lower limit to the integral, π/L, related to the finite size of the system ($L \propto \sqrt{A}$) and an upper limit, π/a due to the finite molecular size (proportional to a), one has the result that

$$\langle h^2(\vec{\rho}) \rangle = \frac{T}{2\pi\gamma} \log \frac{L}{a} \tag{3.18}$$

By symmetry, $\langle h^2(\vec{\rho}) \rangle$ is independent of the point $\vec{\rho}$ in the $x-y$ plane at which the fluctuation is calculated. It is important to note that the mean-square fluctuation diverges as the logarithm of the system size. For a one-dimensional interface (*i.e.*, a line instead of a surface) this divergence is even more severe and increases linearly with the size of the system. (For the line, everything is similar, except that the dimensionality of the integral in Eq. (3.17) is one-dimensional.) Thus, due to reduced dimensionality of these interfaces, the thermal fluctuations can have drastic effects on how "flat" the interfaces really are.

Correlation Functions

The fluctuations also affect the height-height correlations along the surface. Consider the *difference* in heights of two points a distance $\vec{\rho}$ apart (again, $\vec{\rho}$ reflects this distance in the xy plane). Define

$$\Delta^2(\vec{\rho}) = \left\langle (h(\vec{\rho}) - h(0))^2 \right\rangle \tag{3.19a}$$

which is calculated from

$$\left\langle (h(\vec{\rho}) - h(0))^2 \right\rangle = \frac{2}{A} \sum_{\vec{q}} (1 - \cos(\vec{q} \cdot \vec{\rho})) \left\langle |h(\vec{q})|^2 \right\rangle \tag{3.19b}$$

For simplicity, we have chosen one of the points at $\vec{\rho} = 0$ Again, using Eq. (3.16), one can show that $\Delta^2(\vec{\rho})$ diverges as $\log \rho$; as the distance between the two points increases, the difference between their heights increases logarithmically. In the one-dimensional case of an interface which is a line, this divergence is even more severe and is linear in the distance between the points.

Physical Divergences

In the case of a fluctuating surface, these logarithmic divergences are relatively weak: from Eq. (3.18) we see that for typical values of γ of order 100 dyne/cm, $a = 3\text{Å}$, a surface with $L = 100\text{Å}$ has a root-mean-square (rms) fluctuation of about 1.5Å, while one with $L = 1$ cm has an rms fluctuation of only about 7.5Å. In addition, these divergences can be cut off by either the size of system as demonstrated previously, or by other physical effects. In the case of a surface fluctuating in the $\hat{z}$ direction which represents, for example, the interface between a fluid in equilibrium with its vapor or the interface between a light and heavier fluid, gravity (acting in the $\hat{z}$ direction) will tend to suppress the thermal fluctuations, since fluctuations will cost gravitational energy. The gravitational energy of a particle at a height $h(x, y)$ above the plane $z = h = 0$, which measures the zero of energy, is mgh where m is the mass and g is the gravitational constant. For a continuous medium one must integrate this over the density of all the particles from $z = 0$ to $z = h$. Thus, the gravitational energy per unit area is $\int_0^h \rho_0 g z dz = \rho_0 g h^2/2$ where ρ_0 is the density (or density difference for the case of an interface between two fluids). The total free energy of the interface can be written:

$$\Delta F_s = \tfrac{1}{2}\gamma \int dx dy \left[(h_x^2 + h_y^2) + \xi_g^{-2} h^2 \right] \tag{3.20}$$

where $\xi_g^2 = \gamma/\rho_0 g$. The probability distribution of the fluctuations is still Gaussian, and one can repeat the previous calculations with the additional new term proportional to $\xi_g^{-2} \sum_{\vec{q}} |h_{\vec{q}}|^2$. The correlation function is now given by

$$\langle |h(\vec{q})|^2 \rangle = \frac{T}{\gamma(q^2 + \xi_g^{-2})} \tag{3.21}$$

Calculating the mean-square fluctuation of the interface from Eq. (3.17) one sees that the logarithmic divergences are now cut off by the gravitational length ξ_g which is determined by the competition between the surface tension and gravity. Thus, the height fluctuations now have the approximate form, valid when $\xi_g/a \gg 1$, $\langle h^2(\vec{\rho}) \rangle \propto \log(\xi_g/a)$. The correlation functions attain a finite value proportional to $\log(\xi_g/a)$ for correlations between two points whose distance is much larger than the gravitational length, $\rho \gg \xi_g$.

Example: Capillary Wave Dynamics

Derive the equation of motion of an incompressible fluid surface acting under gravity and surface tension under the assumption of negligible fluid viscosity and find the dispersion relation for the surface (capillary/gravity) waves.

To calculate the equation of motion one must consider the dynamics of the two degrees of freedom of the problem: the *dynamic* position of the surface, $z = h(x, y; t)$ (where t is the time) and the fluid velocity, $\vec{v}$. One writes down the total surface, gravitational, and kinetic energy and uses Lagrange's equations[4] to solve for the dynamics.

The potential energy, F_s, is given by a constant term that comes from the bulk (where density variations are not allowed) free energy and a contribution from the surface tension and gravitational terms:

$$F_s = F_0 + \frac{1}{2} \int dA \left[\rho_0 g h^2 + \gamma |\nabla h|^2 \right] \tag{3.22}$$

where A is the area, ρ_0 is the density, and γ is the surface tension. The kinetic energy, T, is written as the integral over the volume, V, of the fluid:

$$T = \frac{1}{2}\rho_0 \int dV v^2 \tag{3.23}$$

For an incompressible system, $\nabla \cdot \vec{v} = 0$ (see Chapter 1) and in addition, for a conservative system, $\vec{v}$ can be written:

$$\vec{v} = \nabla \psi(\vec{r}) \tag{3.24}$$

where ψ is a scalar.

We first consider the bulk system. Using Eq. (3.24), the kinetic energy has the form:

$$T = \tfrac{1}{2}\rho_0 \int dV |\nabla \psi|^2 \tag{3.25}$$

From the discussion of the Euler-Lagrange equations of Chapter 2, this functional is minimized if ψ obeys

$$\nabla^2 \psi = 0 \tag{3.26}$$

Consider a disturbance with a wavevector $\vec{q}$ in the xy plane (*i.e.*, we assume that $h(x, y; t) = Q(t)e^{i\vec{q}\cdot\vec{\rho}}$); we can write ψ in the form:

$$\psi = \tilde{\psi}(z, t) \, e^{i\vec{q}\cdot\vec{\rho}} \tag{3.27}$$

where $\vec{\rho} = (x, y)$. We look for solutions of Eq. (3.26) that obey the boundary conditions (i) $\psi \to 0$ as $z \to -\infty$ (*i.e.*, no motion in the bulk fluid, far from the interface) and (ii) $v_z = \partial\psi/\partial z = \partial h/\partial t$ at $z = h(x, y; t)$, which is the definition of the surface velocity, whose $\hat{z}$ component is $v_z(x, y, z = h)$, in terms of the displacement of the surface. Using Eq. (3.27) in Eq. (3.26) we find that $\tilde{\psi}$ obeys

$$\frac{\partial^2 \tilde{\psi}}{\partial z^2} = q^2 \tilde{\psi} \tag{3.28}$$

The solutions that obey the boundary conditions are written:

$$\tilde{\psi} = \frac{h_t}{|q|} e^{-|q|(h-z)} \tag{3.29}$$

where $h_t = \partial h/\partial t$.

We now use this solution for ψ to integrate the kinetic energy over the $\hat{z}$ direction up to the surface $z = h(x, y; t)$ and find that T_s, obtained by performing the integral in Eq. (3.23) over z, is given by

$$T_s = \int dA \, \frac{\rho_0 h_t^2}{2|q|} \tag{3.30}$$

Now both the kinetic and potential energies are given in terms of the single degree of freedom of the interface, $h(x, y; t)$. One forms the *surface*

Lagrangian, $L_s = T_s - F_s$, which is now only a function of x and y. The equation of Lagrangian dynamics (equivalent to a statement that the total energy is conserved, see Ref. 4), states that the equation of motion is obtained by minimizing the time integral of the Lagrangian:

$$\frac{\delta F_s}{\delta h} = -\frac{\partial}{\partial t}\frac{\partial T_s}{\partial h_t} \tag{3.31}$$

Here, the term $\delta F_s/\delta h$ signifies the variational derivative, where in general,

$$\frac{\delta F}{\delta h(\vec{\rho})} = \frac{\partial f}{\partial h} - \frac{\partial}{\partial \vec{\rho}_i}\frac{\partial f}{\partial h_{\vec{\rho}_i}} \tag{3.32}$$

where $h_{\vec{\rho}_i} = \partial h/\partial \vec{\rho}_i$ and f is the free energy density. In Eq. (3.32) one sums over the repeated index i and $\vec{\rho}_i = (x, y)$. Using this equation of motion and the expressions for the effective surface kinetic and potential energies, one finds that equating the force per unit area and the acceleration per unit area implies that the interface position obeys

$$\rho_0 g h - \gamma(h_{xx} + h_{yy}) + \frac{\rho_0 \ddot{h}}{q} = 0 \tag{3.33}$$

where $h_x = \partial h/\partial x$. We seek a solution of the form:

$$h(x, y; t) = Q(t)\exp[i\vec{q}\cdot\vec{\rho}] \tag{3.34}$$

This implies that $Q(t)$ obeys

$$\ddot{Q} = -q\left(g + \frac{\gamma q^2}{\rho_0}\right)Q \tag{3.35}$$

so that the capillary waves give a sinusoidal time dependence to Q of the form $Q = Q_0 e^{i\omega t}$, where the dispersion relation for ω is

$$\omega = \sqrt{q\left(g + \frac{\gamma q^2}{\rho_0}\right)} \tag{3.36}$$

This simple calculation is valid in the limit of small viscosities; finite viscosity gives rise to a finite damping of these capillary modes[5]. Experimental measurements of this dispersion relation using light scattering yields an estimate of the interfacial tension γ.   ∎

3.4   CAPILLARY INSTABILITIES OF INTERFACES

The example just given showed how to calculate the mode frequency for capillary waves on the surface of a fluid. In the limit of zero viscosity, gravity and surface tension provide the restoring forces and the thermal fluctuations are the driving forces for these spontaneous undulations of the surface. To treat the case of finite viscosity, the energy methods are insufficient and one must solve the Navier-Stokes equation for the undulated interfaces[5] as outlined in Chapter 1. A general theory for the hydrodynamic stability of surfaces and interfaces is presented in Ref. 5; for small viscosities, the damping of the capillary waves vanishes for long wavelengths, $\lambda = 2\pi/q$, as ηq^2, where η is the viscosity. Since the real part of the mode frequency varies as a smaller power of $\vec{q}$ (see Eq. (3.36)), the damping is negligible compared to the period of oscillation, at least in the long-wavelength limit. Of course, when the damping becomes important (at sufficiently large values of $\vec{q}$), the dispersion relation is drastically changed; there is no wave motion and the interface relaxes[5] to equilibrium (*i.e.,* a flat interface).

In the case of capillary waves, the flat surface or interface is basically stable and the waves are the result of thermal fluctuations. However, there are interface geometries where a simple interface shape — such as a cylindrical tube of one phase in other or the surface of a cylinder of fluid or solid — is intrinsically unstable because there are other geometries of lower surface areas and hence lower interfacial free energies. A simple energetic/thermodynamic argument (see the next example) can be used to show that the free energy of an undulated cylinder with undulations whose wavelength exceeds a critical value proportional to the cylinder radius, is lower than that of the perfect cylinder. These undulations eventually lead to the breakup of the cylinder into spheres — which naturally have a lower surface/volume ratio — and is known as the Rayleigh instability.

While the Rayleigh instability of a cylindrical surface is driven by the surface energy, the flat surface of a thin film can become unstable under the influence of van der Waals forces (see Chapter 5). In this case, the surface tension provides a restoring force which stabilizes short-wavelength undulations. Again, the critical wavelength is determined by a balance of the free energies (see the problems at the end of this chapter).

Example: Instability of Flat and Cylindrical Interfaces

Show that a planar interface, governed only by its surface tension is *stable* with respect to small undulations of the interface, while a cylindrical interface is unstable to long-wavelength undulations.

To determine the stability, we consider the interfacial free energy of Eq. (3.1) and determine whether a small perturbation of the interface raises or lowers

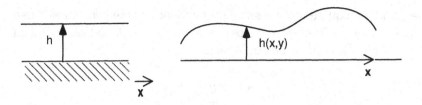

Figure 3.1 Undulations of a flat interface.

the free energy. If the free energy is increased by the perturbation, the surface is stable — although the entropy gained by undulations results in a finite amplitude to these undulations (see Fig. 3.1 and Fig. 3.2) via thermal fluctuations. If the free energy is lowered by the perturbation, the system is inherently unstable and will change its morphology, providing that the kinetic (*e.g.*, hydrodynamic) pathways allow this transformation on an experimentally accessible timescale.

Flat Interface: A flat interface has an interfacial energy given by Eq. (3.1) with $h = 0$. Small perturbations of the interface can be expressed in terms of sinusoidal modes that conserve the volume under the interface, V, given by

$$V = \int dx dy\, h(x,y) = \frac{1}{\sqrt{A}} \sum_{\vec{q}} \int dx dy\, h(\vec{q})\, e^{i\vec{q}\cdot\vec{\rho}} \qquad (3.37)$$

where we have used Eq. (3.13). Expanding Eq. (3.1) for long-wavelength surface perturbations, we find

$$\int dx dy\, \left[1 + \tfrac{1}{2}\left(h_x^2 + h_y^2\right)\right] \geq \int dx dy \qquad (3.38)$$

Thus any such fluctuation in the interface position raises the interfacial area and energy; the flat interface with only interfacial tension is therefore stable.

Cylindrical Interface: Consider a cylinder of unperturbed radius R_0 and length $L \gg R_0$ whose local radius, $\rho(z)$, varies in the $\hat{z}$-direction via

$$\rho(z) = R + \frac{1}{\sqrt{L}} \sum_{q} \rho_q e^{iqz} \qquad (3.39)$$

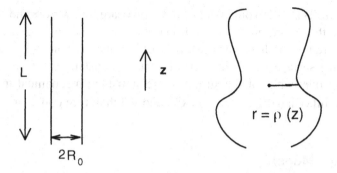

Figure 3.2 Undulations of a cylindrical interface.

This Fourier expansion in terms of the one-dimensional modes propagating in the $\hat{z}$ direction is the most general form for the shape of a deformed cylinder when overhangs are forbidden. In general, the volume of the deformed cylinder is given by

$$V = \pi \int_0^L dz\, \rho^2(z) \tag{3.40}$$

The volume of the unperturbed cylinder (its initial state) is $V_0 = \pi R_0^2 L$. Volume conservation, thus implies

$$R^2 = R_0^2 - \frac{1}{L} \sum_q |\rho_q|^2 \tag{3.41a}$$

If the perturbation is small, $\rho(z) \ll R_0$, and

$$R \approx R_0 \left(1 - \frac{1}{L} \sum_q \frac{|\rho_q|^2}{2R_0^2} \right) \tag{3.41b}$$

The surface free energy of the cylinder is given by

$$F_s = \gamma \int d\theta dz\, \rho(z) \sqrt{1 + \left(\frac{\partial \rho}{\partial z} \right)^2} \tag{3.42}$$

We expand the square root for small perturbations of the cylinder and use Eq. (3.41b) for R. The difference between the total surface free energy of the unperturbed and perturbed cylinder is

$$\Delta F_s \approx \gamma 2\pi R_0 \sum_q \frac{|\rho_q|^2}{2R_0^2} \left(q^2 R_0^2 - 1 \right) \tag{3.43}$$

Thus, the free energy of the perturbed cylinder is *lower* than that of the initial, perfect cylinder when $qR_0 < 1$; long-wavelength perturbations are unstable.

This thermodynamic analysis only indicates that long-wavelength perturbations are unstable. To find the morphology of the new state, one must look at the hydrodynamics of the process that will indicate the fastest growing modes of the system. In addition, the final morphology of the breakup of the cylinder will depend on the effect of the nonlinear terms neglected in this treatment.   ■

3.5   ROUGHENING TRANSITION OF SOLID SURFACES

While simple fluid surfaces show no anisotropy, the macroscopic shapes of crystals are often anisotropic and reflect the underlying symmetry of the crystal lattice. When this happens, it implies that the interfacial fluctuations are insufficient to overcome the energetic anisotropy and one observes sharp facets. However, one might imagine that at sufficiently high temperature (but still below the melting temperature, so that the crystal is still the equilibrium phase), the thermal fluctuations of the interface may overwhelm the anisotropy and the crystal would no longer show sharp facets. The temperature at which this happens is known as the roughening transition temperature; this transition is important in understanding crystal growth[6].

Surface Energy

For a solid surface, where the surface jumps in discrete steps, the continuum energy of Eq. (3.1) may not be accurate enough to account for all the thermodynamic properties. If one considers a fluctuating solid surface described by the height (in the $\hat{z}$ direction) of a "column" located in the xy plane, a more accurate accounting of the extra area is given by the expression

$$\mathcal{H}_c = \tfrac{1}{2} \sum_{ij} J_{ij} |h_i - h_j| \tag{3.44}$$

where J_{ij} is nonzero only when i and j are nearest neighbors and is a positive energy that represents the additional energy of creating the extra surface (compared to the flat interface). In addition, the variables $\{h_i\}$ are discrete since the surface moves in steps of a lattice constant. Even this expression ignores overhangs of the surface. The statistical mechanical treatment of the Hamiltonian, $\mathcal{H}_c$, is complex[6,7] and we therefore consider a simplified model.

Surface Model

The simplified model, which is a good approximation to Eq. (3.44) when the local jumps in the height between nearest neighbors are limited to either zero or one lattice constant:

$$\mathcal{H} = \int dx dy \left[\tfrac{1}{2} \gamma |\nabla h|^2 + y_0 \left(1 - \cos \frac{2\pi h(x,y)}{a} \right) \right] \qquad (3.45)$$

This approximation replaces the factor $h_i - h_j$ by $(h_i - h_j)^2$, which in the continuum is replaced by the gradient. (The continuum approximation is not crucial and the calculation could also be performed with the discrete differences with similar results.) In addition, the requirement that the height in a solid makes jumps in discrete integral steps of the lattice constant, a, is replaced by an energy proportional to $y_0 > 0$, which has units of energy per unit area and is a minimum when the height is an integral multiple of a. As we shall see, at high temperatures, when the interface shows substantial fluctuations, this energy is irrelevant and the continuum approximation of Eq. (3.1) adequately describes the surface. At very low temperatures, this approximate Hamiltonian is no longer accurate. However, an understanding of the statistical mechanics of a system with the model Hamiltonian of Eq. (3.45) highlights most of the important physics.

Variational Approach

We shall investigate the system using the approximate variational approach described in Chapter 2. We consider a reference Hamiltonian $\mathcal{H}_0$, which is tractable, and use the theorem that bounds the exact free energy F_e, by

$$F_e < F = F_0 + \langle \mathcal{H} - \mathcal{H}_0 \rangle_0 \qquad (3.46)$$

where F_0 is the free energy of the model system; the average values $(\langle \ldots \rangle_0)$ are taken with respect to the Boltzmann factor $e^{-\mathcal{H}_0/T}$ of the reference Hamiltonian. We then vary the parameters of $\mathcal{H}_0$ to get the lowest upper bound on the exact free energy. Since Gaussian integrals are most easily performed, we choose

$$\mathcal{H}_0 = \tfrac{1}{2} T \sum_{\vec{q}} G(\vec{q}) h(\vec{q}) h(-\vec{q}) \tag{3.47}$$

where $h(\vec{q})$ are the Fourier components defined in Eqs. (3.13,3.14). The entire function $G(\vec{q})$ can be regarded as determining the set of "parameters" which characterize the reference Hamiltonian; these parameters (and hence the function $G(\vec{q})$) are determined by the minimization. The free energy of the reference Hamiltonian, F_0, is given by

$$F_0 = -T \log Z_0 = -T \log \left[\prod_{\vec{q}} \int dh(\vec{q}) \, e^{-\mathcal{H}_0/T} \right] \tag{3.48}$$

Performing the Gaussian integral we find

$$F_0 = \tfrac{1}{2} \sum_{\vec{q}} T \log \left(\frac{G(\vec{q})}{2\pi} \right) \tag{3.49}$$

and the average of the reference Hamiltonian is $\langle \mathcal{H}_0 \rangle_0$ is a constant. The average of the gradient square term in $\mathcal{H}$ gives a term proportional to $\sum q^2 \langle |h(\vec{q})|^2 \rangle$ and for the Gaussian probability distribution,

$$\langle |h(\vec{q})|^2 \rangle_0 = 1/G(\vec{q}) \tag{3.50}$$

To perform the average of the term proportional to y_0, we use the fact that

$$\left\langle e^{ikh(r)/a} \right\rangle_0 = e^{-k^2 g_0/2} \tag{3.51}$$

where

$$g_0 = \left(\frac{1}{Aa^2} \right) \sum_q [G(\vec{q})]^{-1} \tag{3.52}$$

and A is the area. Thus, lumping all the constants into a term F_c, we find

$$F = F_c + \tfrac{1}{2} \sum_{\vec{q}} T \log G(\vec{q}) + \tfrac{1}{2} \gamma \sum_{\vec{q}} \frac{q^2}{G(\vec{q})} - A y_0 e^{-2\pi^2 g_0} \tag{3.53}$$

Minimization of the Free Energy

We now minimize F with respect to $G(\vec{q})$ and find the equation that defines $G(\vec{q})$

$$\frac{T}{G(\vec{q})} - \gamma \frac{q^2}{G(\vec{q})^2} - \frac{4\pi^2 y_0 a^{-2}}{G(\vec{q})^2} e^{-2\pi^2 g_0} = 0 \qquad (3.54)$$

We thus find

$$G(\vec{q})^{-1} = \frac{T}{\gamma(q^2 + \xi^{-2})} \qquad (3.55)$$

where

$$\xi^{-2} = \frac{4\pi^2 y_0}{\gamma a^2} e^{-2\pi^2 g_0} \qquad (3.56)$$

Since g_0 is related to a sum over all the $G(\vec{q})$, Eq. (3.55) has to be solved self-consistently to yield a value for ξ. Thus, performing the sum over $\vec{q}$ in Eq. (3.55) we obtain g_0, which is then used in Eq. (3.56) to derive an equation for ξ. The result is

$$g_0 = \frac{T}{4\pi\gamma a^2} \log\left[1 + \left(\frac{\pi\xi}{a}\right)^2\right] \qquad (3.57)$$

Since ξ is defined by an exponential of g_0, the logarithm contributes to a power law dependence as a function of $\tau = T\pi/(2\gamma a^2)$ and

$$\left(\frac{a}{\pi\xi}\right)^2 = \frac{4y_0}{\gamma}\left[1 + \left(\frac{\pi\xi}{a}\right)^2\right]^{-\tau} \qquad (3.58)$$

Roughening Transitions

The solution of these equations is shown in Fig. 3.3 for the case where $4y_0/\gamma = 1$ for different values of τ. The general trend for the solution is that when $\tau > 1$ (high temperatures) there is only one solution at $\xi^{-1} = 0$. Thus, at high temperatures, the effects of the lattice are negligible and the system is well approximated by the continuum surface-tension Hamiltonian since $G(\vec{q}) \propto \gamma q^2$. One can also show that the average of the term proportional to y_0 is zero in this regime. One says that the surface is "rough" when $\tau > 1$, meaning that the effects of the lattice are irrelevant. The correlation functions show divergent behavior as discussed in Section 3.3 for the case of the system with only surface tension. As the temperature is lowered ($\tau < 1$)

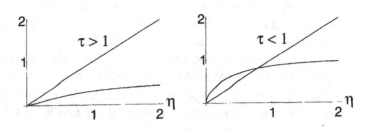

Figure 3.3 Graphical solution of Eq. (3.58) for the variable $\eta = \left[a/(\pi\xi)\right]^2$. We have plotted the results for the case where $4y_0/\gamma = 1$.

Equation (3.58) has two solutions: one at $\xi^{-2} = 0$ and another at ξ^{-2} finite. By explicitly finding the free energy, one can show that the finite value of ξ is the solution that has the lower free energy. The surface is said to be "smooth" since the correlation function has the form $1/(q^2 + \xi^{-2})$. As shown for the case of gravity as a restoring force at the end of Section 3.3, the length ξ now acts as a cutoff for the fluctuations, which no longer diverge. The position of the surface as well as the difference in the height-height correlation function are now finite and the surface is well defined. The point where $\tau = 1$ is defined as the "roughening transition" temperature and the variational approximation presented here correctly predicts this temperature. However, a more precise treatment[6,7], shows that the correlation length at the roughening transition temperature is a nonanalytic function of $\tau - 1$. As one approaches the roughening temperature from below ($\tau < 1$), ξ diverges as

$$\xi \sim \exp\left[\frac{c_0}{(1-\tau)^{1/2}}\right] \tag{3.59}$$

where c_0 is a constant.

Example: Correlation Length for Roughening Transition

The nature of the solutions of the self-consistent expression, Eq. (3.58), can be examined by considering the variable $\eta = \left[a/(\pi\xi)\right]^2$. In this case, Eq. (3.58) becomes

$$\eta = f(\eta) = \frac{\alpha}{\left(1 + \frac{1}{\eta}\right)^\tau} \tag{3.60}$$

where $\alpha = 4y_0/\gamma$. In a graphical solution of this equation for η one would look for the crossings of η and $f(\eta)$ plotted versus the variable η. The first plot is naturally a straight line with a slope of unity. The second plot has zero slope at infinity, independent of the value of τ as shown in Fig. 3.3. For values of $\tau > 1$, the slope of $f(\eta)$ in the limit of $\eta \to 0$ tends to zero and the two curves do not cross as long as $\alpha < 1$. For large values of α, a crossing is possible, but the value of η determined by this crossing would be large, implying a correlation length smaller than the microscopic cutoff; this is not a physical result. Thus the only physical solution is an infinite correlation length as for the surface undulations of a fluid. For $\tau < 1$, the slope of $f(\eta)$ for small values of η tends to infinity, guaranteeing a crossing of the two curves and hence a finite correlation length. These conclusions are correct so long as the periodic potential can be treated perturbatively ($\alpha < 1$); the resulting values of $\eta \neq 0$ are small, consistent with our assumptions.   ■

Effect on Dynamics of Growth

Whether the crystal surface is smooth or rough can have important consequences for the dynamics of crystal growth[8]. Above the roughening transition, when the fluctuations are large, there are no energetic barriers to crystal growth. Thus, if one puts a driving force, such as a temperature gradient, on the system, which will cause the solid phase to grow at the expense of the liquid phase, the solid will grow with a rate proportional to the driving force. The resulting crystal surface will be smooth and not faceted. At low temperatures, the effects of the lattice constant show up in that the driving force has to exceed a critical value for the crystal surface to grow; this is because the interface grows in steps in a macroscopically uniform manner below the roughening transition — *i.e.,* one must move an entire plane of atoms together. Above the transition, the thermal fluctuations allow the atoms to move more independently.

3.6 PROBLEMS

1. Equilibrium Crystal Shapes

Summarize the effect of the roughening transition on the equilibrium shapes of crystals as discussed in C. Jayaprakash, W. F. Saam, and S. Teitel, *Phys. Rev. Lett.* **50**, 2017 (1983) and C. Rottman and M. Wortis, *Phys. Rev. B* **29**, 328 (1984); *Phys. Repts.* **103**, 59 (1984).

2. Surface Correlation Functions

To find the complete expressions for the height-height correlation functions in real space (Eq. (3.19)), one has to evaluate the integrals:

$$\int d\vec{q}\,\frac{(1 - \cos \vec{q} \cdot \vec{r})}{q^2 + \xi^{-2}} \tag{3.61}$$

Consider this integral in both one (line interface) and two dimensions (surface interface) with the upper cutoff set to ∞. Use them to write down more accurate expressions for the correlation functions and the mean square fluctuation of the surface. (For the two-dimensional case, use integral tables for integrals of Bessel functions. Express the result in terms of another Bessel function that you should also expand for small values of its argument.)

3. Hydrodynamics of Capillary Waves

Derive the dispersion relation for capillary waves on the surface of a fluid with nonzero viscosity. Since this is a dissipative system, the energy approach used in the text is not applicable and one must start from the hydrodynamic equations for an incompressible fluid. Assume that that amplitude of the wave and its velocity are small, linearize the hydrodynamic equations and calculate both the real and imaginary parts of the dispersion relation for these (possibly overdamped) waves.

4. Rupture of Thin Films

Thin films can be unstable to undulations of their surfaces when the van der Waals energy (see Chapter 5) is such that the thin film has lower energy than a thick film. Consider a thin layer of initial thickness D_0, whose free energy is composed of a surface tension contribution as well as a van der Waals contribution F_v for the energy per unit area, which varies with the *local* film thickness, $h(x, y)$, as

$$F_v = -\frac{A}{12\pi h^2(x, y)} \tag{3.62}$$

Here, $A > 0$ is the "Hamaker constant", which has units of energy and characterizes the strength of the interaction (see Chapter 5). Determine the conditions under which the free energy of an undulated layer is lower than that of the flat layer.

5. Roughening Free Energies

Using the general form of the correlation function of Eq. (3.55), calculate the free energy of the fluctuating interface of a solid as a function of ξ and show that at high temperatures there is only one solution corresponding to $\xi \to \infty$ and that at low temperatures there are two solutions, but that the free energy for finite ξ is lower than that of infinite ξ.

6. Roughening in One Dimension

Show that the equilibrium roughness of a one-dimensional interface is qualitatively larger than that of a two-dimensional interface. Show that at least for small values of y_0, that the one-dimensional interface is always rough (*i.e.*, the roughening transition temperature is zero).

3.7   REFERENCES

1. K. Kawasaki and T. Ohta, *Prog. Theoretical Phys.* **67**, 147 (1982).

2. E. H. Lucassen-Reynders and J. Lucassen, *Adv. Coll. Interf. Sci.* **2**, 347 (1969).

3. See Part B — Interfacial Waves in *Physiochemical Hydrodynamics*, ed. M. G. Velarde (Plenum Press, New York, 1988), p. 147.

4. H. Goldstein, *Classical Mechanics* (Addison-Wesley, Reading, MA, 1950).

5. C. A. Miller and P. Neogi, *Interfacial Phenomena* (Marcel Dekker, New York, 1985).

6. J. D. Weeks in *Ordering in Strongly Fluctuating Condensed Matter Systems*, ed. T. Riste (Plenum Press, New York, 1980), p. 293.

7. M. E. Fisher in *Statistical Mechanics of Membranes and Surfaces*, eds. D. Nelson, T. Piran, and S. Weinberg (World Scientific, Teaneck, NJ, 1989), p. 19.

8. Y. Saito, *Z. Phys. B* **32**, 75 (1978).

CHAPTER **4**

Wetting of Interfaces

4.1 INTRODUCTION

In the previous chapter, we described the energetics, fluctuations, and phase transitions of surfaces and interfaces. The surface or interface is the mathematical idealization of the region separating two coexisting phases. In this chapter, we move to the next level of complexity and treat both the statics and dynamics of three coexisting phases in the context of the topic of wetting[1,2]. For simplicity, we consider the case of a solid substrate in equilibrium with both a fluid and vapor and determine the structure of the fluid layer. The treatment is easily generalized to the problem of a solid substrate supporting a phase-separated binary mixture as well as to the problem of fluid substrates[3]. We first treat the problem in the classical manner and show how a simple analysis of the competing interfacial tensions determines the contact angle of the fluid drop on the solid. We next show how the classical theory is related to a more microscopic description of the problem and in particular, we use the theory of the profile of the liquid layer on the solid substrate to relate the contact angle that the macroscopic drop makes with the solid, to the thickness of a thin layer of fluid that covers the substrate in the region between the drops. Finally, we discuss the dynamics of wetting

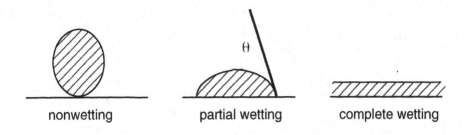

nonwetting   partial wetting   complete wetting

Figure 4.1 Depiction of a fluid drop atop a solid substrate. Complete wetting is characterized by a contact angle $\theta = 0$, partial wetting by $0 < \theta < \pi$, and nonwetting by
$\theta = \pi$.

since many experimental situations involve nonequilibrium effects, and mention
recent work on pattern formation in wetting.

4.2   EQUILIBRIUM: MACROSCOPIC DESCRIPTION

Complete and Partial Wetting

We begin by considering a solid substrate covered by a fluid drop that is in
equilibrium with its vapor. For simplicity, we first consider a perfect substrate
(a perfectly smooth solid surface) and a very dilute vapor that is approximated
by a vacuum. When the drop of fluid is put on top of a solid surface there
are two competing effects. The interactions with the solid substrate may make it
energetically favorable for the drop to spread so that it wets the surface. However,
spreading increases the area of contact between the fluid and the vapor. This
increases the surface tension energy between the drop and the vapor and can
destabilize the wetting film. When the interaction with the solid surface (*e.g.,*
favoring the liquid-solid interface over the vapor-solid interface) dominates, one
gets complete wetting and when the surface tension term dominates, one gets
partial wetting as sketched in Fig. 4.1.

For partial wetting, one defines a contact angle, θ which is given, in equilibrium, by a balance of macroscopic forces. That is, one considers the change
in energy by moving the contact line by a small distance; force balance in equilibrium requires that the energy be stationary with respect to this displacement.
We define the energies per unit area: γ_{sl} between the solid and liquid drop,
γ_{sv} between the solid and the vapor or vacuum, and γ between the liquid drop
and the vapor (*i.e.,* the surface tension of the liquid drop). Consider an infinite,
wedge-shaped region that represents the intersection of the drop very close to

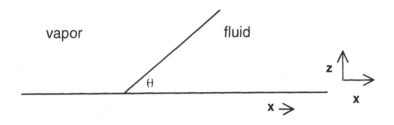

Figure 4.2 A wedge of fluid atop a solid substrate. The contact angle is θ.

the substrate (whose normal is in the $\hat{z}$ direction) at a contact angle θ, as shown in Fig. 4.2. Displacing the contact line a distance dx in the $\hat{x}$ direction exposes an area $L dx$ to the vapor where L is the size of the system in the $\hat{y}$ direction, with a corresponding free energy cost of $(\gamma_{sv} - \gamma_{sl}) L dx$, since the exposed area no longer has liquid-solid contact. In addition, the liquid-vapor interface has its area decreased by an amount $L \cos \theta dx$. Thus the change in surface free energy, dF is given by

$$dF = L \left(\gamma_{sv} - \gamma_{sl} - \gamma \cos \theta \right) dx \tag{4.1}$$

Requiring $dF/dx = 0$ is equivalent to balancing the forces in the $\hat{x}$ direction, and we find

$$\cos \theta = \frac{\gamma_{sv} - \gamma_{sl}}{\gamma} \tag{4.2}$$

This is the famous Young equation[4].

Increasing either the solid-liquid surface energy, γ_{sl}, or the liquid-vapor interfacial tension, γ, tends to favor partial wetting, while increasing the solid-vapor surface energy, γ_{sv}, tends to favor complete wetting since the solid surface prefers to be wet by the liquid. When $\gamma_{sl} - \gamma_{sv} < \gamma$ but $\gamma_{sl} - \gamma_{sv} > 0$, the surface tension dominates and there is only partial wetting with a finite contact angle that is larger than $\pi/2$. When $\gamma_{sv} > \gamma_{sl}$, but $\gamma_{sv} - \gamma_{sl} < \gamma$, the contact angle is nonzero and less than $\pi/2$. As the surface tension, γ, is decreased so that $(\gamma_{sv} - \gamma_{sl})$ and γ become equal, there is complete wetting of the substrate by the drop. In equilibrium, nothing else can happen since if $\gamma_{sl} + \gamma < \gamma_{sv}$, there would always be a liquid film and never a true equilibrium between the solid and vapor; *i.e.*, the solid-vapor system would build up a macroscopically thick fluid layer between them, thus causing complete wetting. Note that Young's law assumes that the contact angle is determined by the macroscopically measurable surface tensions. However, in the case of long-range interactions that diverge as the film thickness decreases — *e.g.*, van der Waals interactions — the energies of contact can be modified. The following section shows how complete wetting can be prevented to some degree by such interactions.

In addition, experimental observations of the contact angle in some systems may not reflect equilibrium; surface inhomogeneities can give rise to contact angle hysteresis which can modify the applicability of Young's law to dynamical measurements of the contact angle. Contact angle hysteresis refers to the fact that when the contact angle is measured dynamically, one can obtain different values of the contact angle depending on whether one is measuring an advancing (*i.e.*, increasing the solid-liquid contact area) or receding (decreasing the solid-liquid contact area) contact line[1]; these deviations in the contact angle are often associated with impurities on the surface which pin the contact line and change the local contact angle. When the contact line moves with a finite velocity, it may not have time to equilibrate due to these pinning effects and the measured contact angle reflects these deviations. The equilibrium contact angle must then be carefully defined as the limit of the measured values as the velocity is decreased to zero.

Example: Droplet Shape and Contact Angle

Young's equation for the contact angle can also be derived from a full variational treatment of the shape of a finite-sized droplet (and not just a semi-infinite wedge) on a solid surface. Derive this relationship in the approximation of small contact angles for a one-dimensional geometry (with the variation of the profile in the $\hat{x}$ direction, $h(x)$). The drop extent in the $\hat{x}$ direction is parameterized by λ, which must be found self-consistently using the constraint of volume conservation. The extension to larger contact angles and/or three-dimensional drops is straightforward.

The free energy of the drop on the substrate (in the small slope approximation), relative to the free energy of a vapor-covered substrate is

$$f = \int_{-\lambda}^{\lambda} dx \left[\gamma_{sl} - \gamma_{sv} + \gamma \left(1 + \tfrac{1}{2} \left(\frac{\partial h}{\partial x} \right)^2 \right) \right] \qquad (4.3)$$

since the drop extends from $-\lambda < x < \lambda$. The free energy is minimized subject to the constraint that the total cross-sectional area is conserved. One thus minimizes

$$g = f - \mu \int_{-\lambda}^{\lambda} dx \, h(x) \qquad (4.4)$$

where the Lagrange multiplier, μ, is determined from the conservation law,

$$\int_{-\lambda}^{\lambda} dx \, h(x) = A_0 \tag{4.5}$$

where A_0 is the constant cross-sectional area.

For the case of small contact angles considered here, we can derive Young's law (Eq. (4.2)), without explicitly solving for the drop profile. To do this, one considers the constrained free energy, g of Eq. (4.4) as a function of two independent variables, $h(x)$ and λ. Now, the minimization over $h(x)$ requires fixing $h = 0$ at $x = \lambda$ so that these two variables are not completely independent. However, if we rescale the x coordinate and define $u = x/\lambda$, the boundary conditions read:

$$h(u = -1) = h(u = 1) = 0 \tag{4.6}$$

This symmetry implies $\partial h(u)/\partial u = 0$ at $u = 0$. With this scaling, we can take the derivatives with respect to h and λ independently. We define the spreading power, S by

$$S = -(\gamma + \gamma_{sl} - \gamma_{sv}) \tag{4.7}$$

so that $S > 0$ indicates a tendency to complete wetting and $S < 0$ results in partial wetting. We then have, using Eqs. (4.4,4.7):

$$g = \lambda \left[\int_{-1}^{1} du \left(-S + \frac{\gamma}{2\lambda^2} \left(\frac{\partial h(u)}{\partial u} \right)^2 - \mu h(u) \right) \right] \tag{4.8}$$

Functionally minimizing g with respect to $h(u)$ yields

$$\frac{\gamma}{\lambda^2} \frac{\partial^2 h(u)}{\partial u^2} = -\mu \tag{4.9}$$

whose first integral is

$$\frac{\gamma}{2\lambda^2} \left(\frac{\partial h(u)}{\partial u} \right)^2 = -\mu h(u) + C \tag{4.10}$$

where C is a constant. This integration constant can be determined from the minimization of g with respect to λ, where one must consider the derivatives with respect to the overall factor of λ as well as the derivative of the terms involving λ in the gradient expression:

$$\int_{-1}^{1} du \left(-S - \frac{\gamma}{2\lambda^2} \left(\frac{\partial h(u)}{\partial u} \right)^2 - \mu h(u) \right) = 0 \qquad (4.11)$$

(Note the negative sign in front of the gradient term.) Using Eq. (4.10) in Eq. (4.11) determines $C = -S$ so that evaluating Eq. (4.10) at $u = \pm 1$, where $h = 0$, determines the contact angle at the boundary

$$\theta = \left(\frac{\partial h}{\partial x} \right)_{x=-\lambda} = \sqrt{\frac{-2S}{\gamma}} \qquad (4.12)$$

This is equivalent to Young's equation in the limit of small contact angles.

We can also derive Young's equation by explicitly minimizing the grand potential of Eq. (4.4) with respect to the droplet shape. The resulting Euler-Lagrange equation is

$$-\gamma \frac{\partial^2 h}{\partial x^2} = \mu \qquad (4.13)$$

which has the parabolic solution

$$h = \frac{\mu}{2\gamma}(\lambda^2 - x^2) \qquad (4.14)$$

This solution obeys the boundary conditions that the drop meet the substrate at $x = \pm\lambda$: $h(x = \pm\lambda) = 0$. The exact solution is a section of a circle (two dimensions) or section of a sphere (three dimensions); it here approximated by a parabola for small slopes. The free energy corresponding to this solution is

$$f = 2 \left[\lambda(\gamma + \gamma_{sl} - \gamma_{sv}) + \frac{\mu^2\lambda^3}{6\gamma} \right] \qquad (4.15)$$

Now, the Lagrange multiplier, μ, is related to the extent of the drop, λ, by Eq. (4.5), which implies

$$\frac{2\mu\lambda^3}{3\gamma} = A_0 \qquad (4.16)$$

We now minimize f with respect to the extent of the drop, λ, and find

$$\frac{1}{2}\frac{\partial f}{\partial \lambda} = (\gamma + \gamma_{sl} - \gamma_{sv}) + \frac{\mu^2 \lambda^2}{2\gamma} + \frac{\mu \lambda^3}{3\gamma}\left(\frac{\partial \mu}{\partial \lambda}\right) = 0 \qquad (4.17)$$

From Eq. (4.16), we determine

$$\frac{\partial \mu}{\partial \lambda} = -\frac{3\mu}{\lambda} \qquad (4.18)$$

so that Eq. (4.17) implies

$$(\gamma + \gamma_{sl} - \gamma_{sv}) - \frac{\mu^2 \lambda^2}{2\gamma} = 0 \qquad (4.19)$$

But, the contact angle

$$\theta = \left(\frac{\partial h}{\partial x}\right)_{x=-\lambda} = \frac{\mu \lambda}{\gamma} \qquad (4.20)$$

Thus, Eqs. (4.19,4.20) imply

$$\theta = \sqrt{\frac{-2S}{\gamma}} \qquad (4.21)$$

where the spreading coefficient, S, is defined by Eq. (4.7).

Note that the case of small contact angles treated here means that $S \approx 0$, which only occurs when $\gamma_{sl} - \gamma_{sv} < 0$. This is the same expression for the contact angle as Eq. (4.2), for small contact angles, when $\cos \theta \approx 1 - (\theta^2/2)$. Although the macroscopic derivation of Young's equation is simpler, the approach described here, which self-consistently solves for the profile, can be generalized to explicitly include the interactions between the fluid and vapor and the substrate as described below. ∎

4.3   LONG-RANGE INTERACTIONS: MACROSCOPIC THEORY

Long-Range Forces

Up to now, we considered the competition between the surface tension and the fluid-substrate interactions in determining whether a system will partially or completely wet a solid substrate. In the case of complete wetting, the thickness of the fluid layer in between the solid and vapor is only determined by the amount of material in the system. A finite amount of fluid on an infinitely large substrate will result in a monolayer of fluid wetting the surface. However, the presence of long-range forces can result in a minimum to the free energy at a particular value of the film thickness. A recent discussion of several possibilities for the drop shape in the presence of van der Waals interactions is presented in Ref. 5; here we consider the simplest case of complete wetting on an infinitely large substrate.

van der Waals Energy

To estimate this optimal thickness, consider the van der Waals energy of a fluid film. As we shall discuss in Chapter 5, van der Waals energies arise from polarization fluctuations in the medium and occur in all materials. The microscopic van der Waals energy between two molecules, arising from the induced dipole-induced dipole interaction varies as $1/r^6$ in the static approximation (where the finite time for the propagation of light is negligible, applicable for distances less than the order of $800\,\text{Å}$); for thick films, the interaction falls off faster than $1/r^6$. Integrating the nonretarded interaction over all pairs of molecules in a film of thickness h gives an interaction energy,

$$V(h) = \frac{A}{12\pi h^2} \tag{4.22}$$

for the nonretarded case. The sign of the coefficient A (the Hamaker constant), is determined by the dielectric functions of both the fluid and the substrate (see Chapter 5). We consider the case where the coefficient $A > 0$, which tends to thicken the film. This competes with the surface interactions, which tend to thin the film to produce complete coverage of the surface.

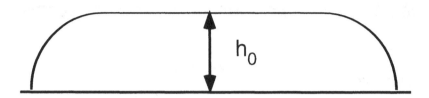

Figure 4.3 Pancake-like droplet formed when one takes into account the thickening van der Waals interactions. The maximum height, h_0, is a function of both the Hamaker constant and the interfacial energies as discussed in the text.

Wetting Profile: Complete Wetting

We now consider the case of complete wetting (spreading power $S \geq 0$) for the case where there are van der Waals interactions, which tend to thicken the film. For a finite amount of fluid spreading on an infinite solid substrate, the equilibrium film profile will therefore not be a monolayer, but a "pancake" with a maximum thickness that is determined by the balance of the surface tensions and the van der Waals energies (see Fig. 4.3).

For simplicity, imagine a two-dimensional drop (with a one-dimensional interface between the fluid drop and the vapor). We assume that the substrate area is larger or equal to the size of the drop and that it therefore does not constrain the droplet profile. The free energy, f, of the drop on the substrate is given by

$$f = \int_{-L}^{L} dx \left[-S + \tfrac{1}{2}\gamma \left(\frac{\partial h}{\partial x} \right)^2 + V[h(x)] \right] \tag{4.23}$$

where we have used the expression for the surface tension for small contact angles. We have also assumed that the van der Waals energy is related to the *local* thickness of the film, since we use the expression for a uniform film with a spatially dependent thickness, $h(x)$. This approximation is valid in the limit where the thickness varies slowly in space. The drop extends from $x = -L$ to $x = L$, and S includes the energy of interaction with the substrate as well as the energy to make a flat interface: $S = \gamma_{sv} - \gamma_{sl} - \gamma$ using the notation defined previously. We consider the minimization of f subject to the constraint of a fixed total amount of material (this will also determine L). Thus we minimize

$$g = f - \mu \int_{-L}^{L} dx \, h(x) \tag{4.24}$$

The variation of g with respect to the profile $h(x)$ is given by the Euler-Lagrange equation:

$$\gamma\frac{\partial^2 h}{\partial x^2} = -\Pi[h] - \mu \tag{4.25}$$

where $\Pi[h] = -\partial V/\partial h$ and is known as the disjoining pressure. Multiplying Eq. (4.25) by $\partial h/\partial x$ and integrating we find

$$\frac{\gamma}{2}\left(\frac{\partial h}{\partial x}\right)^2 = V[h] - \mu h + C \tag{4.26}$$

In Eq. (4.26), the constant of integration, C can be shown to equal $(-S)$ by a procedure similar to that discussed following Eq. (4.10). At the midpoint, $x = 0$, the profile is symmetric; defining $h(0) = h_0$ we obtain, $V(h_0) - \mu h_0 - S = 0$, which gives μ as a function of h_0. The actual profile can be found by integrating Eq. (4.26) to give x/λ as a function of h/h_0. The relevant scale for x, is the distance $\lambda = h_0^2/(\sqrt{3}a)$, where a is a length defined by the competition between the van der Waals energy and the surface tension:

$$a = \sqrt{\frac{A}{6\pi\gamma}} \tag{4.27}$$

and is of order 3 Å for typical materials. To find the equilibrium height h_0, one can use Eq. (4.25) and the expression for μ in terms of V and S. Anticipating that at $x = 0$, the curvature, $\partial^2 h/\partial x^2 \approx 0$ (it vanishes as $L \to \infty$), we find

$$h_0 = a\sqrt{\frac{3\gamma}{2S}} = \sqrt{\frac{A}{4\pi S}} \tag{4.28}$$

depends only on A and S, since it arises from a balance of the van der Waals and contact energies. Recalling that large values of $S > 0$ indicate an increased spreading tendency, we see that decreasing S or increasing A or a tends to thicken the film. Thus this calculation shows that there is a preferred value of h_0; this is relevant when $L \gg \lambda$ since otherwise the conservation constraint determines the profile.

4.4   FLUCTUATIONS OF THE CONTACT LINE

In many problems of real-world interest, substrates are not infinitely smooth nor are they chemically homogeneous. Disorder in either the structure or the chemistry of the substrate will result in local changes in the contact angle and hence the shape of the wetting profile. Instead of a partially wetting, three-dimensional wedge meeting the substrate at a well-defined contact angle, the locus of this intersection being a straight line, the three-phase contact line will tend to wander in space due to the disorder. This occurs because both the local position and local contact angle fluctuate on a surface with a variable value of the surface energy; the interface "finds" the optimal position and angle on the surface which lowers its free energy. These nonidealities can also profoundly affect the dynamics of spreading by giving rise to local equilibria that pin the contact line and result in contact angle hysteresis (see Ref. 1).

Here, we focus on the static effects of inhomogeneities on the contact line and the wedge profile. We estimate the loss of correlations (*i.e.*, the wigglyness) in the position of the contact line for a fluid drop contacting a substrate with a chemically inhomogeneous surface[1,6] so that the solid-vapor and solid-liquid tensions differ from their average by a random function $w(x, y)$ whose correlations are given by $\langle w(x, y) \rangle = 0$ (*i.e.*, the mean fluctuation in the tension is zero) and

$$\langle w(x, y)w(x', y') \rangle = w_0^2 e^{-|\vec{r}|/\xi} \tag{4.29}$$

where $\vec{r}^2 = (x - x')^2 + (y - y')^2$ and ξ is the correlation length for the surface impurities.

Consider a wedge of fluid as in Fig. 4.4, where the contact line for the perfect substrate is a straight line in the $\hat{y}$ direction given by $x = 0$. The surface tensions for contact with the perfect solid are $\gamma_{sv}^{(0)}$ and $\gamma_{sl}^{(0)}$ for contact with the vapor and liquid respectively and the equilibrium contact angle is θ_0. For the inhomogeneous substrate, the contact line is given by $x = X(y)$ (see Fig. 4.4) and the contact energy (*i.e.*, interaction with the solid substrate) of the system which is covered by vapor in the region $-\infty < x < X(y)$ and by liquid in the region $X(y) < x < \infty$, is

$$F_s = \int dy \int_{-\infty}^{X(y)} dx \, \gamma_{sv} + \int dy \int_{X(y)}^{\infty} dx \, \gamma_{sl} \tag{4.30}$$

The integrals over y extend over the entire system, which has length in the $\hat{y}$ direction that we denote as L. The difference in energy between this system and

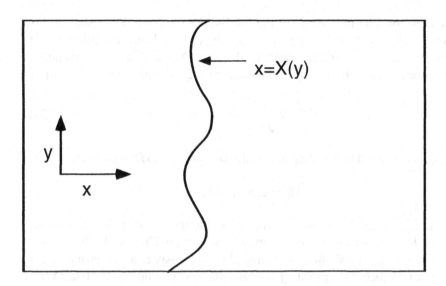

Figure 4.4 Top view of a wedge of fluid with a contact line that meets the surface along a boundary that varies in the $\hat{y}$ direction as $x = X(y)$. The region to the right of the boundary is covered by fluid (which has locally wedgelike shape in the $\hat{z}$ direction), and the region to the left of the boundary is covered by the vapor.

the contact energy of the perfect solid where the contact line is also at $x = x(y)$ is

$$\Delta F_s = \int dy \int_0^{X(y)} dx \, w(x, y) \tag{4.31}$$

where

$$w(x, y) = \left[\gamma_{sv} - \gamma_{sl} - \left(\gamma_{sv}^{(0)} - \gamma_{sl}^{(0)} \right) \right] \tag{4.32}$$

and where we assume that the average of $w(x, y)$ over y is zero. We have neglected constant terms and we have used the fact that the average fluctuation in the solid-vapor and solid-liquid tensions is zero. Treating the disorder as a perturbation around the case of the clean surface with a contact line at $x = 0$, we approximate the value of the surface energy disorder at the contact line, $x = X(y)$, by its value at $x = 0$. We thus set $x = 0$ in the argument of $w(x, y)$ and approximate

$$\Delta F_s \approx \int dy \, X(y) \, w(0, y) \tag{4.33}$$

We now consider the energy of the fluid-vapor interface. The shape of this interface is deformed by the fluctuations of the contact line. The difference in energy of this interface described by a height, $h(x, y)$, above the inhomogeneous substrate compared to the energy of the perfect wedge-shaped region is given by

$$\Delta F_v = \tfrac{1}{2}\gamma \int dy \int_{X(y)}^{\infty} dx \left[h_x^2 + h_y^2 - \theta_0^2 \right] \tag{4.34}$$

where $h_x = \partial h / \partial x$. The total change in the free energy, ΔF is given by

$$\Delta F = \Delta F_s + \Delta F_v \tag{4.35}$$

We must minimize this energy with respect to the two degrees of freedom, $h(x, y)$ and $X(y)$, which determine the shape and position of the fluctuating interface respectively. We first minimize ΔF with respect to the shape of the fluid-vapor interface, $h(x, y)$ for a *fixed* but spatially varying contact line, $X(y)$. Thus, $\delta \Delta F / \delta h(x, y) = 0$ implies

$$h_{xx} + h_{yy} = 0 \tag{4.36}$$

with the boundary conditions (i) $h(X(y), y) = 0$ and (ii) as $x \to \infty$, $h(x, y)$ approaches its unperturbed value. Fourier transforming both $X(y)$ and $h(x, y)$ in the $\hat{y}$ direction with wavevector q and using Eq. (4.36) gives a dependence of $h(x, q) \sim \exp(-|q|x)$ for each Fourier component. Summing over all Fourier components and choosing the constants so that the boundary conditions are satisfied, keeping terms to first order in $X(y)$ (*i.e.*, assuming $|q|X(q) \ll 1$), we find the solution:

$$h(x, y) = \theta_0 \left[x - \frac{1}{\sqrt{L}} \sum_q X(q) e^{iqy} e^{-|q|x} \right] \tag{4.37}$$

where L is the size of the system in the $\hat{y}$ direction. At large values of x, this expression for the profile is equivalent to the unperturbed wedge. At small values of $x = X(y) \approx 0$, one can approximate $e^{-|q|x} \approx 1$ (since the expression for the profile is already first-order in $X(y)$), and one sees that the boundary condition $h = 0$ when $x = X(y)$ is also obeyed.

Now that we have minimized to find the liquid-vapor interface shape at fixed contact line position, $X(y)$, we evaluate the free energy for this shape (Eq. (4.37)) and minimize to find the optimal value of $X(y)$. We therefore put Eq. (4.37) in Eq. (4.34) for the capillary energy and integrate over the x-coordinate from $x =$

$X(y) \approx 0$ (since there is already a term proportional to $X(y)$ in the expression for $h(x,y)$) to find the change in the interface free energy, ΔF_v, as

$$\Delta F_v = \frac{\gamma \theta_0^2}{2} \sum_q |q| \, |X(q)|^2 \qquad (4.38)$$

Fourier transforming Eq. (4.33) and defining $w(q)$ as the Fourier transform of $w(0,y)$ in the $\hat{y}$ direction, we write the change in the surface free energy, ΔF_s, as

$$\Delta F_s = \sum_q w(-q)X(q) \qquad (4.39)$$

Minimizing the total free energy per unit length, $\Delta F = \Delta F_s + \Delta F_v$ with respect to the Fourier transform of the contact line coordinate, $X(q)$, we find

$$X(q) = -\frac{w(q)}{|q|\gamma\theta_0^2} \qquad (4.40)$$

The interface position adjusts to the surface inhomogeneity to minimize its total free energy.

By analogy with the description of roughening given in Chapter 3, the correlation function describing the wandering of the position of the contact line over a distance y is given by

$$\Delta^2(y) = \left\langle (X(y) - X(0))^2 \right\rangle = \frac{2}{L} \sum_q (1 - \cos qy) \left\langle |X(q)|^2 \right\rangle \qquad (4.41a)$$

where the averages here are not over the thermal fluctuations, but rather over the randomness due to the surface impurities. We thus find

$$\Delta^2(y) = \frac{2}{L\gamma^2\theta_0^4} \sum_q (1 - \cos qy) \frac{\left\langle |w(q)|^2 \right\rangle}{q^2} \qquad (4.41b)$$

If $w(x,y)$ has the spatial correlations described in Eq. (4.29), then a simple scaling analysis indicates that

$$\Delta^2(y) \sim \frac{w_0^2}{\gamma^2\theta_0^4} \xi y \qquad (4.42)$$

Note that although the fluctuations of the surface energy have a characteristic length scale given by ξ, the mean-square interface roughness is *not* proportional to ξ^2, but to a much longer range function, ξy, which depends on the distance between the two points. This is because the deformation of the vapor-fluid interface by the fluctuations in the contact line induce an effective long-range interaction between different points along the contact line. Thus, the root-mean-square fluctuation of the interface position varies with the square root of the distance between the two points. This variation is similar to the way that the *thermal* roughness of the contact line induces fluctuations in its position.

4.5   EQUILIBRIUM: MICROSCOPIC DESCRIPTION

Solid-Liquid-Vapor Coexistence

We have seen that complete wetting is characterized by a macroscopic liquid film in between the solid substrate and the vapor. Macroscopically, partial wetting is characterized by the solid substrate covered by a wedge or finite drop of the fluid. In the macroscopic picture, the remainder of the solid contacts the vapor. This is all that can be said from the macroscopic point of view, that considers only two possible states: the liquid and vapor. However, interactions with the solid can result in a local density of molecules near the substrate that is quite different from the density of either the liquid or the vapor. We consider the following cases: (i) The interaction with the solid substrate is repulsive, favoring a *low density* of molecules on the substrate — *i.e.,* a gas. In this case, there is solid-vapor equilibrium and one would not expect small contact angle wetting of the substrate by the liquid. (ii) The interaction with the substrate is attractive, favoring a *high density* of the fluid. In this case, there is the possibility for solid-fluid-vapor equilibrium, with a layer of fluid on the substrate whose density is larger than that of the vapor, but may not equal that of the bulk fluid phase. When this intervening fluid layer is macroscopically thick, one has complete wetting. When this intervening fluid layer is not macroscopically thick, it can coexist with a macroscopic liquid phase. A self-consistent treatment of both phases predicts both the thickness of the fluidlike layer as well as the contact angle of the bulk wedge or drop of fluid.

Wetting Model: Uniform Coverage

We first consider the case where the substrate is uniformly covered by a layer of "fluid", but that this layer is not necessarily at the equilibrium fluid density. In general, the layer will not be homogeneous in density as one approaches the vapor — *i.e.,* if the normal to the substrate is in the $\hat{z}$ direction, the layer density will vary with $\hat{z}$. Our goal is first to find the relationship between the thickness of the layer and the interactions of the fluid molecules with the substrate. We consider the free energy per unit area, f_s, of the fluid on the surface to be given by a virial expansion of the form

$$f_s[n_s] = \left[a_0 - a_1 \frac{n_s}{(n_\ell - n_v)} + \tfrac{1}{2} a_2 \frac{n_s^2}{(n_\ell - n_v)^2} ... \right] \tag{4.43}$$

where $n(\vec{r})$ is the density of the fluid-vapor above the solid ($z > 0$), $n_s = n(x, y, z = 0)$ is the density of the fluid at the solid surface, and n_ℓ and n_v are the bulk, equilibrium liquid and vapor densities respectively. In Eq. (4.43), we have normalized the expansion by $(n_\ell - n_v)$ (which is fixed by the bulk free energy) so that the coefficients of n_s all have the dimensions of an interfacial tension (energy per unit area). The first term in this expression is a constant term, the second represents an attraction (when $a_1 > 0$) of the molecules to the surface (recall that if the molecules are repelled from the surface it is highly unlikely for a fluidlike layer to exist near the surface of the substrate), and the third term represents an "excluded-volume" type of interaction (which can be repulsive or attractive) of the molecules adsorbed on the surface. For simplicity, we consider the case where this interaction is repulsive ($a_2 > 0$) so the added complication of a liquid-gas phase separation by the adsorbed molecules is absent.

The total interfacial free energy per unit area, f_t, consists of the sum of f_0 and the free energy per unit area that comes from the liquid-vapor interface. In equilibrium, one minimizes the total free energy subject to the conservation constraint — i.e., one works at fixed chemical potential. As explained in the discussion of the gas-liquid interface in Chapter 2, the appropriate bulk free energy to minimize to find the interfacial profile is the grand potential per unit area, g_s, which is written:

$$g_s = \int dz \left[W[n(z)] + \tfrac{1}{2} B |\nabla n(z)|^2 \right] \tag{4.44}$$

where W includes the chemical potential and also subtracts off the grand potential of the uniform bulk phase. In Eq. (4.44), we have specialized to the case of a one-dimensional profile, and $g_s = G_s/A$ is a free energy per unit area. The expression for g_s contains both the gradient energy of the interface, as well as the difference between the local free energy at equilibrium of the bulk gas or liquid and the local free energy of an arbitrary density variation. For gas-liquid equilibrium, we have seen in Chapter 2 that

$$W = c(n - n_v)^2 (n - n_\ell)^2 \tag{4.45}$$

where n_v and n_ℓ are the equilibrium gas and liquid densities and c is a constant, related to the critical density, which has units of the product of an energy and a volume cubed.

Minimization of the Free Energy

We consider the total grand potential per unit area, g_t, given by

$$g_t = f_s[n_s] + g_s \qquad (4.46)$$

which is parameterized by the spatially varying density, $n(z)$, and by the value of the density on the surface, $n_s = n(z = 0)$. We thus first minimize g_t with respect to the $n(z)$ with a *fixed* value of the surface density, n_s to find the profile for arbitrary surface densities. This depends only on g_s. Afterward, one uses this solution for $n(z)$ to minimize g_t again to find the local surface density, including the contributions from both f_s and g_s, both of which depend on n_s.

The Euler-Lagrange equation arising from the minimization of g_s with respect to the density profile, $n(z)$, is

$$B\frac{\partial^2 n}{\partial z^2} = \frac{\partial W}{\partial n} \qquad (4.47)$$

Multiplying both sides by $\partial n/\partial z$ one finds that

$$\tfrac{1}{2}B\left(\frac{\partial n}{\partial z}\right)^2 = W[n(z)] \qquad (4.48)$$

where we used the boundary condition that $W = 0$ in either the liquid or vapor phases where the gradient terms vanish. Using Eq. (4.48) back in the expression for the free energy, we write:

$$g_t = f_s[n_s] + \int_{n_b}^{n_s} dn\sqrt{2WB} \qquad (4.49)$$

where n_b is the bulk value of the density (either liquid or vapor).

We now find the surface density, n_s, by minimizing Eq. (4.49) with respect to n_s. This can give two different types of solutions: (i) $n_s \ll n_\ell$, indicating that the surface layer is a gas and that there is essentially a solid-vapor interface with no intervening liquid, wetting film, and (ii) $n_s \approx n_\ell$, indicating that there is a fluid wetting film on the solid with a density comparable to that of the bulk liquid. In this case, one can either have one solution to $\partial g_t/\partial n_s = 0$, which indicates complete wetting by a fluid layer. If there are two (stable) solutions to this equation, we have partial wetting; the two solutions correspond to a thin film in contact with the substrate, coexisting with a wedge of the fluid phase. (The surface concentration in the thin-film region is lower than the surface concentration in the wedge.) The relationship of the contact angle of this macroscopic region to the thickness of the thin film is examined next.

Thin-Film Profile

The calculation of the profile is similar to that of the bulk gas-liquid interface, discussed in Chapter 2. We consider the case where $n_v < n_s < n_\ell$. From Eq. (4.48) with the form of W given by Eq. (4.45) one has

$$\frac{\partial n}{\partial z} = \pm\sqrt{\frac{2c}{B}}\,(n - n_\ell)(n - n_v) \qquad (4.50)$$

Defining

$$\Delta = \frac{n_s - n_v}{n_\ell - n_s} > 0 \qquad (4.51)$$

we can solve Eq. (4.50) with the boundary condition $n = n_s$ at $z = 0$ (this is where this situation differs from the bulk coexistence of Chapter 2) and find

$$n(z) = \frac{(n_v - n_\ell) + n_\ell(1 + \Delta e^{-z/\xi})}{1 + \Delta e^{-z/\xi}} \qquad (4.52)$$

where now,

$$\xi = \sqrt{\frac{B}{2c}}(n_\ell - n_v)^{-1} \qquad (4.53)$$

If $n_s \gg n_v$, then

$$n(z) = n_\ell\left[e^{(z-z_0)/\xi} + 1\right]^{-1} \qquad (4.54)$$

where

$$z_0 = \xi \log\left(\frac{n_s}{n_\ell - n_s}\right) \qquad (4.55)$$

A schematic plot of this function is shown in Fig. 4.5. There is a thin film with an almost constant density $n_s \approx n_\ell$, of thickness z_0, after which the density drops and approaches that of the vapor phase. The thickness, z_0 is directly related to the surface density and increases logarithmically as $n_s \to n_\ell$. It is determined

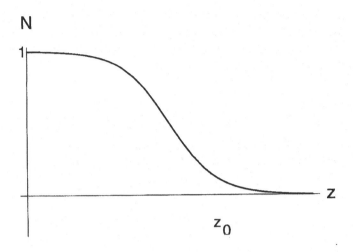

Figure 4.5 Interfacial profile for gas-liquid system with a solid substrate where $N = n/n_\ell$ (the density scaled to the equilibrium liquid density) and $Z = z/\xi$ as defined in the text. The curve is calculated for the case where $n_v \ll n_\ell$.

by the surface interactions and the bulk density and correlation length and is *not* macroscopic (*i.e.*, does not scale with the system size).

One can observe an increase in the thickness of the film, z_0, by changing temperature. We consider the case where n_s is relatively insensitive to temperature, but as is usual for the liquid-gas transition, the liquid density, $n_\ell(T)$, decreases as one approaches the bulk transition temperature for the phase separation. One then expects a finite thickness film of size z_0 at low temperatures where $n_\ell \gg n_s$ and a transition to an infinitely thick layer at some higher temperature, T_w, when $n_\ell(T_w) = n_s$. We show below that this film (with a finite value of z_0) *coexists* with both the vapor and the bulk fluid phase, with the bulk fluid phase in the shape of a wedge that meets the substrate at an equilibrium contact angle related to z_0. (This is somewhat different from usual bulk coexistence where one considers regions with no particular geometrical relationship.) As the temperature approaches the wetting transition temperature, T_w, $n_\ell(T_w) \to n_s$, and the thickness of the film diverges; there is a macroscopically thick fluid layer in between the solid and vapor — *i.e.*, a transition to complete wetting. This transition is not necessarily a continuous one as shown in the following discussion.

Wetting Transitions in the Film Thickness

We now determine the equilibrium value of the density on the surface and hence the equilibrium film thickness — again for the case where the film is uniform in the $\hat{x}$ direction. Using Eq. (4.45) for the local grand potential, W, and performing the integration in Eq. (4.49), we find that the total free energy per unit area is

$$g_t = f_s + \gamma \left[1 - 3\tilde{\psi}^2 + 2\tilde{\psi}^3\right] \tag{4.56}$$

where

$$\tilde{\psi} = \frac{(n_\ell - n_s)}{(n_\ell - n_v)} \tag{4.57}$$

and

$$\gamma = \int_{n_v}^{n_\ell} dn\sqrt{2WB} \tag{4.58}$$

is precisely the vapor-liquid interfacial energy defined in Chapter 2 — a quantity that is independent of n_s. One can also expand the first three terms in Eq. (4.43) in the small quantity $\tilde{\psi}$ to yield

$$f_s[n_s] = f_s[n_\ell] + \left[a_1 - \frac{a_2\,n_\ell}{(n_\ell - n_v)}\right]\tilde{\psi} + \frac{a_2}{2}\tilde{\psi}^2 \tag{4.59}$$

where $f_0[n_\ell]$ is independent of n_s and depends only on the bulk liquid density.

For relatively thick films, the quantity $\tilde{\psi}$ is small. From Eq. (4.55) we can write:

$$\tilde{\psi} = \frac{\psi}{(1 + \psi)} \tag{4.60}$$

where

$$\psi = \exp(-z_0/\xi) \tag{4.61}$$

Combining Eqs. (4.56,4.59,4.60), we can expand for small values of ψ which corresponds to large local thicknesses, z_0 and find

$$g_t = f_s[n_\ell] + U[\psi] \tag{4.62a}$$

where

$$U[\psi] \approx \gamma + \alpha\psi + \tfrac{1}{2}\beta\psi^2 + \tfrac{1}{3}\Delta\psi^3 \tag{4.62b}$$

U

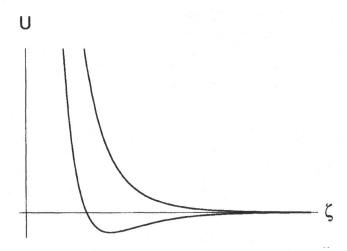

ζ

Figure 4.6 Plot of the local free energy, U, as a function of $\zeta = z_0/\xi$, the local, dimensionless height of the film. This plot is for the case $\beta > 0$ and represents a second-order wetting transition for the thickness of the film; as α goes through zero the minimum at finite ζ moves continuously to infinity.

Here,

$$\alpha = a_1 - a_2' \tag{4.63a}$$

where $a_2' = a_2 n_\ell/(n_\ell - n_v)$ and is approximately equal to a_2 far from the critical point. We also find that

$$\beta = -2(a_1 - a_2') + a_2 - 6\gamma \tag{4.63b}$$

$$\Delta = 3(a_1 - a_2) - 3a_2' + 24\gamma \tag{4.63c}$$

Depending on the relationship of the parameters a_1, a_2, a_2', and γ, the coefficients α, β, and Δ can have varying signs and the minimization of the local part of the free energy, $U[\psi]$, with respect to the local thickness, related to ψ by Eq. (4.61) can have the characteristics of either a second-order transition (when $\alpha \approx 0$, $\beta > 0$) or a first-order transition (when $\alpha \approx 0$, $\beta < 0$, and $\Delta > 0$) (see Fig. 4.6 and Fig. 4.7).

Thus, the equilibrium film thickness can either be finite (the lowest minimum of U with respect to ψ occurs at a finite value of z_0 and ψ) or infinite (the lowest minimum of U with respect to ψ occurs at $z_0 \to \infty$ or $\psi = 0$). This transition can occur continuously or with a discontinuous jump in the value of z_0 as α or the

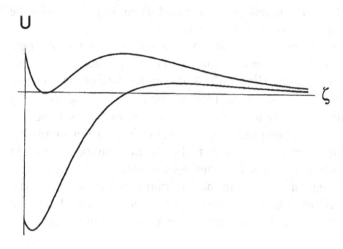

Figure 4.7 Plot of the local free energy, U, as a function of $\zeta = z_0/\xi$, the local, dimensionless height of the film. This plot is for the case $\beta < 0$ and represents a first-order wetting transition for the thickness of the film, for small values of α; the minimum at finite values of ζ jumps discontinuously to a global minimum at $\zeta \to \infty$ when α reaches a critical value.

temperature is varied. For more general surface interactions (*e.g.*, those that might arise from long-range van der Waals interactions), one generally expects U to show a first-order transition, since it will consist of very-short-range interactions with the substrate (terms that are powers of ψ) as well as longer-range interactions that will vary as inverse powers of z_0 (see the problems at the end of this chapter).

Coexistence of Thin Film and Fluid Wedge: Partial Wetting

The previous discussion assumed that the film density varied in the $\hat{z}$ direction only and derived a characteristic length scale, z_0, which relates the film thickness to the density on the substrate and to the equilibrium liquid density. However, partial wetting is most often associated with the finite contact angle of the macroscopic liquid drop that coexists with this thin film. We now show that the contact angle is related to the equilibrium film thickness and compute the entire profile of the two coexisting phases. The reason for this coexistence is that when partial wetting is preferred, the system tends to form a thin film with a particular thickness. However, the requirement that there be a macroscopic amount of fluid implies that this thin film must coexist with a macroscopic wedge of fluid. The fluid profile — the transition from thin film to drop — relates the contact angle of the wedge to the film thickness.

In order to examine the coexistence of a wedge with a finite contact angle and the thin film, we have to consider the variation of the profile and hence of the local value of what was termed z_0 (Eqs. (4.54,4.55)) in the plane of the substrate (in the $\hat{x}$ and $\hat{y}$ directions). We therefore identify the *local* value of z_0 with a profile height, $h(x)$ and reserve the notation z_0 to denote the thickness of the thin film in equilibrium — from the minimization of Eqs. (4.60,4.62). For simplicity, we focus on a one-dimensional variation of the height, $h(x)$, in the $\hat{x}$ direction for the geometry shown in Fig. 4.8, but consider the *continuous* variation of the profile from the wedge at $x \to \infty$ to the flat, thin film as $x \to -\infty$. The free energy has a contribution from the fluid-vapor surface tension, due to the variation in the $\hat{x}$ direction, which contributes a free energy per unit area:

$$f_a[h(x)] = \frac{\gamma}{2} \int dx \left(\frac{\partial h(x)}{\partial x} \right)^2 \tag{4.64}$$

There is also a contribution from the combination of the interactions with the substrate and the gas-liquid interfacial energy due to the variation of the *density* in the $\hat{z}$ direction; this is contained in g_t of Eq. (4.62) and we consider only the part that depends on $h(x)$, namely $U[\psi(h(x))]$ defined in Eq. (4.62). We thus have a profile with a free energy per unit area, f_p:

$$f_p = f_a[h(x)] + \int dx\, U[\psi[h(x)]] \tag{4.65}$$

where ψ and h are related as in Eq. (4.61):

$$\psi = \exp\left[-\frac{h(x)}{\xi} \right] \tag{4.66}$$

The functional minimization of f_p with respect to the profile, $h(x)$, along with the boundary condition that $\partial h / \partial x \to 0$ when ψ or $h(x)$ approach their equilibrium (thin-film) values of ψ_0 and z_0 respectively, yields

$$\tfrac{1}{2}\gamma \left(\frac{\partial h}{\partial x} \right)^2 = U[\psi(h(x))] - U_0 \tag{4.67}$$

where $U_0 = U[\psi_0]$ (and $\psi_0 = \exp[-z_0/\xi]$) is the value of U at the absolute minimum of $U[\psi]$. This value of the thickness corresponds to the equilibrium value, z_0.

From Eq. (4.67) we can immediately relate the equilibrium contact angle, θ_0, of the macroscopic wedge where $h \to \infty$ and $\psi \to 0$, as determined by

$$\theta_0 = \left(\frac{\partial h}{\partial x}\right)_{\psi=0} = \sqrt{\frac{2(U_\infty - U_0)}{\gamma}} \qquad (4.68)$$

where in our model, $U_\infty = U(\psi[h \to \infty]) = \gamma$. This is directly related to the height of the thin film where ψ is nonnegligible by the dependence of U_0 on ψ_0. Eq. (4.67) can be integrated to explicitly yield the profile consisting of both the wedge (*e.g.*, at large positive values of x) and the thin film (*e.g.*, at large negative values of x).

In the particular case where we have a second-order wetting transition, an analytical form for the profile can be derived. We consider Eq. (4.62) with α small and negative, $\beta > 0$ and neglect the term proportional to Δ since we focus on small values of ψ and the free energy is already stabilized by $\beta > 0$. The equilibrium value of the film thickness is thus given by $z_0 = -\xi \log(|\alpha|/\beta)$ (this is equivalent to Eq. (4.55)) and the minimization of Eq. (4.65) with respect to $h(x)$ yields

$$h(x) - z_0 = \xi \log\left(1 + e^{\kappa x}\right) \qquad (4.69)$$

where $\kappa = |\alpha|/\sqrt{\beta\gamma}$. This profile, shown in Fig. 4.8, extrapolates smoothly between the thin film with asymptotic thickness, z_0, to the wedge, at large values of x.

The equilibrium contact angle of the wedge, defined by the x-derivative of $h(x)$ for large values of x is therefore given by $\theta_0 = \kappa\xi$ and vanishes when $\alpha = 0$ and z_0 diverges. We note that the contact angle, related to κ, vanishes linearly as $\alpha \to 0$ and the wetting transition is approached, while the thickness of the thin film with which the macroscopic drop is in equilibrium diverges only logarithmically as α vanishes. This calculation explicitly demonstrates the relationship between the vanishing of the equilibrium contact angle and the divergence (albeit logarithmic) of the thickness of the film that coexists with the fluid wedge. Comparing this expression for the contact angle with the macroscopic statement of Young's law, Eqs. (4.2,4.21), we see that the theory has fixed the values of the spreading coefficient (the solid-liquid and solid-vapor interfacial energies) in terms of the microscopic interactions. A similar analysis can be performed for first-order wetting transitions where it can be shown that at the first-order transition, the contact angle does vanish; however, this happens when the film thickness z_0 is finite and jumps in a first-order manner to an infinite value and complete spreading.

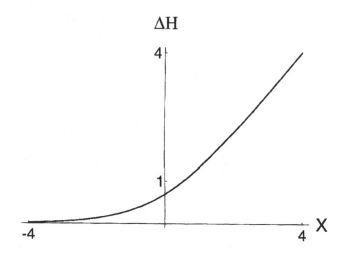

Figure 4.8 The profile of the wedge and the thin film determined by the theory, for a system with a second-order wetting transition. Here $\Delta H = (h(x) - z_0)/\xi$ so that $\Delta H = 0$ is the top of the thin film whose thickness is z_0. In the figure, we define $X = \kappa x$. In these units, the contact angle in the figure is always finite, but when one takes the scalings into account, it is seen that the observable contact angle is equal to κ, which vanishes at the transition.

4.6   DYNAMICS OF WETTING

Spreading Dynamics

In many practical problems it is of interest to understand how the spreading process develops as a function of time. For example, one may prepare a drop of fluid with a given initial contact angle that is not necessarily equal to the equilibrium contact angle governing the fluid-substrate interface. The time dependence of approach of the contact angle to its equilibrium value determines how quickly the spreading process will occur. As we shall see, for simple situations, the approach to complete wetting is very slow. A further area of interest is in the patterns that are formed as fluids spread on solids (*e.g.,* paint drips, nonuniformity of spin coatings of lubricants on magnetic disks). These patterns are intimately related to *dynamical* instabilities of thin, wetting films.

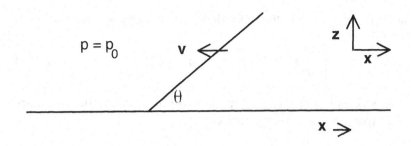

Figure 4.9 A wedge of fluid moving along a solid substrate with velocity $\vec{v}$ in the $-\hat{x}$ direction. The dynamic contact angle is denoted by θ, which varies with time. We consider the case where the equilibrium contact angle is zero.

Hydrodynamics of Wetting

We shall consider the spreading of a macroscopic wedge of fluid on top of a preexisting thin layer of fluid (*e.g.*, which arises from the van der Waals interactions) as shown in Fig. 4.9. The seemingly simpler problem of the spreading of a macroscopic drop of fluid across a completely dry substrate is complicated by the fact that the usual *stick boundary condition*, which governs the solid-fluid interface does not allow the fluid right near the solid to advance and wet since this boundary condition requires that the velocity of the fluid vanishes at the solid interface. One way around this problem is to assume that in some region near the solid, the stick boundary condition does not apply. Another is to study the spreading atop a preexisting fluid film. We have seen that such thin films are stabilized by the van der Waals interactions. Wetting dynamics can be derived[7] self-consistently, taking into account the van der Waals interactions in a fundamental manner. For simplicity, we imagine that even for complete wetting, the thickness of the thin layer at infinity is fixed (and nonzero) due to the van der Waals interactions and consider spreading on top of a thin layer of fixed height, b.

Hydrodynamic Equations

We consider a fluid of a uniform density, ρ, moving with velocity $\vec{v}(\vec{r}, t)$. The conservation of mass or continuity equation implies

$$\frac{\partial \rho}{\partial t} + \nabla \cdot (\rho \vec{v}) = 0 \tag{4.70}$$

and the balance of forces within the fluid (the Navier-Stokes equation) states:

$$\rho \left[\frac{\partial \vec{v}}{\partial t} + (\vec{v} \cdot \nabla)\vec{v} \right] = -\nabla p + \eta \nabla^2 \vec{v} \tag{4.71}$$

where η is the viscosity and p is the pressure. We also require the boundary condition of no forces $f_{s,i}$ on the free surface:

$$f_{s,i} = -p\, n_i + \eta \frac{\partial v_k}{\partial x_i}\, n_k = 0 \tag{4.72}$$

where n_i is the ith component of the normal unit vector and one sums on the repeated index (k). The term proportional to the viscosity is the frictional force on the surface. On the solid surface, we have the stick boundary condition, $\vec{v} = 0$.

Lubrication Approximation

For a thin, incompressible film (thin compared with the variations in the profile in the $\hat{x}$ and $\hat{y}$ directions), we assume that in the Navier-Stokes equations, $\vec{v}$ is only z dependent; thus,

$$\nabla^2 \vec{v} \to \frac{\partial^2 \vec{v}}{\partial z^2} \tag{4.73}$$

In addition, we assume that since the film is thin, the pressure is constant in the $\hat{z}$ direction. Thus the only components of ∇p are in the $\hat{x}$ and $\hat{y}$ directions. Finally, we consider steady-state solutions of the Navier-Stokes equation: $\partial \vec{v}/\partial t = 0$. Writing $\vec{v} = \vec{v}(z)$, these equations are

$$\eta \frac{\partial^2 v_x}{\partial z^2} = \frac{\partial p}{\partial x} \tag{4.74a}$$

$$\eta \frac{\partial^2 v_y}{\partial z^2} = \frac{\partial p}{\partial y} \tag{4.74b}$$

Since we take p to be independent of z within the lubrication approximation, $v_z = 0$. The pressure drop across a curved interface located at $z = h(x, y)$ is given by the Laplace condition (see Chapter 2):

$$p = p_0 - \gamma(h_{xx} + h_{yy}) \tag{4.75}$$

where $h_x = \partial h/\partial x$, and the boundary conditions on the velocity are $v(0) = 0$ and $\partial \vec{v}/\partial z = 0$ at $z = h(x, y)$. With these boundary conditions, Eq. (4.74) is solved as

$$\eta \vec{v}(z) = \tfrac{1}{2} \nabla p(x, y) \left[z^2 - 2zh \right] \qquad (4.76)$$

where p is independent of z and is given by Eq. (4.75). The average velocity, $\vec{U}$, is determined from

$$\vec{U} = \frac{1}{h} \int_0^h \vec{v}(z)dz = \frac{\gamma}{3\eta} h^2 \, \nabla_\perp \left(\nabla_\perp^2 h \right) \qquad (4.77)$$

In Eq. (4.77),

$$\nabla_\perp = \hat{x} \frac{\partial}{\partial x} + \hat{y} \frac{\partial}{\partial y} \qquad (4.78)$$

The characteristic velocity of the fluid is of the order $\gamma/\eta \approx 10^3$ cm/s, while the contact line velocity is much smaller since it depends on the curvature which is small; typically $U \approx 1$ cm/s.

Wetting Profile

The profile (see Fig. 4.9) is determined from the conservation law, Eq. (4.70). Integrating this equation from $z = 0$ to $z = h(x, y)$ yields

$$\frac{\partial h}{\partial t} + \nabla_\perp \cdot h\vec{U} = 0 \qquad (4.79)$$

Using the expression for the average velocity in Eq. (4.77) we obtain the time dependence of the profile from the solution of

$$\frac{\partial h}{\partial t} + \nabla_\perp \cdot \frac{\gamma}{3\eta} h^3 \, \nabla_\perp \left(\nabla_\perp^2 h \right) = 0 \qquad (4.80)$$

The profile obeys the boundary condition that $h = b$ (the height of the pre-existing fluid film) for distances far from the wetting front (i.e., as $x \to -\infty$).

One-Dimensional Solution

We now limit ourselves to the case where the profile, $h(x)$, varies only in the $\hat{x}$ direction. In this case, Eq. (4.80) reads:

$$\frac{\partial h}{\partial t} + \frac{\partial}{\partial x}\left[\frac{\gamma}{3\eta}h^3 h_{xxx}\right] = 0 \qquad (4.81)$$

If the wetting front moves with a velocity, $U_0(t)$, that has a "slow" time dependence, we can look for a solution where $h = h(x + \int dt U_0)$, but with no other explicit time dependence. (This approach is justified by our final result.) Thus, in forming $\partial h/\partial t$ we keep only the term $U_0 h_x$ which we will find to be much larger than the term $(dU_0/dt)h$. We then find

$$\frac{\partial}{\partial x}\left[\text{Ca } h + \tfrac{1}{3}h^3 h_{xxx}\right] = 0 \qquad (4.82)$$

where the capillary number, Ca is defined as $\text{Ca} = \eta U_0/\gamma$; we note that $\text{Ca} \ll 1$ according to the preceding discussion. Using the boundary condition that $h = b$ when we are far from the wetting front and there is only the flat, thin film where all derivatives of h vanish, we have

$$3\,\text{Ca}\,(h - b) + h^3 h_{xxx} = 0 \qquad (4.83a)$$

In the region of the macroscopic drop (*i.e.*, large, positive values of x), $h \gg b$ and

$$3\,\text{Ca}\,h + h^3 h_{xxx} = 0 \qquad (4.83b)$$

This equation has the approximate solution

$$h \approx x\left[3\,\text{Ca}\,\log\left(\frac{x}{x_0}\right)\right]^{1/3} \qquad (4.84)$$

where x_0 is obtained by matching this solution to that of the thin film, and is of the order of b/θ, where the contact angle θ is determined from $\theta = (\partial h/\partial x)$.

Droplet Spreading Dynamics

This solution for the profile is almost a wedge with a constant slope. The dynamic contact angle, θ is

$$\theta = \left[3\,\mathrm{Ca}\,\log\left(\frac{x}{x_0}\right)\right]^{1/3} \tag{4.85}$$

By the definition of the capillary number, we therefore find that $\theta \propto U_0^{1/3}$; slow velocities imply small dynamic contact angles. This expression[8] has been termed **Tanner's law**.

Example: Growth Law of Spreading Drops

Calculate the time dependence of the growth of a three-dimensional droplet using Tanner's law and conservation of matter.

We consider (three-dimensional) droplet with a finite volume whose wetting front is spreading according to the theory outlined above. The drop size, R, obeys $R\theta \approx h$ and the conservation of volume implies that $R^2 h \approx \Omega_0^{1/3}$, where Ω_0 is a constant. Using the fact that $\theta \propto U_0^{1/3}$ and $dR/dt = U_0$, we find that the drop size obeys the law:

$$R^9 \frac{dR}{dt} = \tilde{\Omega}_0 \tag{4.86}$$

where $\tilde{\Omega}_0$ is another constant. This equation indicates that the drop grows very slowly as a function of time: $R \propto t^{1/10}$ and the average velocity $U_0 \propto t^{-9/10}$ is small as claimed earlier. For a two-dimensional droplet, a similar calculation of the spreading dynamics yields $R \propto t^{1/7}$. This growth law still represents very slow dynamics of this system. ∎

Forced Wetting and Pattern Formation

Since the time development of spontaneous wetting is so slow ($R \propto t^{1/10}$), external forces are often used to speed up the wetting process. For example, a drop can be placed on a rotating turntable, which thins and spreads the drop by centrifugal force[9] or fluid can be made to flow down an inclined plane[10]. Temperature gradients can also speed up the spreading of drops[11]. The external force results in spreading that is much faster than the $t^{1/10}$ law; for gravity forcing, for example, the front moves as $t^{1/3}$. However, as the experiments show, such spreading results in a profile that is flat far from the contact line. Near the contact line, the profile is given by Tanner's law and in between these two regions, the profile develops a "hump" with an excess of fluid behind the contact line. For a three-dimensional system, this hump is similar to a cylinder of fluid that is unstable[12] to fingering in a manner similar to the Rayleigh instability of a cylinder discussed in Chapter 3. Thus, such driven systems do not spread uniformly, but rather through fingers of a definite width that move across the turntable or down the inclined plane.

4.7 PROBLEMS

1. Fluctuating Contact Line: Random Defects

Why does the effective capillary energy for the wiggly contact line vary linearly with q (Eq. (4.38)) when a naive estimate based on the scaling of interfacial tension would have predicted a q^2 dependence? Relate this to a general property of Laplace's equation that the wedge height satisfies.

Calculate the correlation function in Eq. (4.41) by performing the integrals.

2. Fluctuating Contact Line: Periodic Defects

Consider the case where the substrate is perturbed in a periodic manner in the $\hat{y}$ direction. In a region of size D in the vicinity of each defect, the substrate is perturbed with a strength w_0 and is unperturbed outside that region. Using Eq. (4.38) and the appropriate representation of the defect free energy, calculate the displacement of the contact line at $x = 0$.

3. Partial Wetting Profile with van der Waals Interactions

Consider the profile of a wedgelike fluid drop which contacts a solid substrate. In addition to the interfacial energies, include the van der Waals energy $\sim A/h^2$, where $A > 0$ indicates a preference for thickening the film, and h is the local film thickness. Minimize the free energy to find the profile for the case where h is much larger than the thickness of the thin film on the solid with which the wedge coexists. The results should therefore be independent of the detailed surface interactions and just depend on the van der Waals energy and the equilibrium contact angle.

Can you show that the combination of a short-range interaction with the substrate and the van der Waals interaction result in a first-order wetting transition? [*Hint*: Expand the free energy for small values of ψ or relatively large values of h.]

4. Complete Wetting and van der Waals Interactions

For the case of complete wetting, a van der Waals interaction with $A > 0$ tends to thicken the film. Consider the case of a finite drop of fluid on a substrate using the macroscopic equations for the free energy (*e.g.*, Eq. (4.26) with the chemical potential given by Eq. (4.25) evaluated at $h = h_0$ where the curvature of the drop can be taken to be zero) and calculate the equilibrium profile.

5. Profile for First-Order Wetting Transition

By calculating the profile of both the wedge of macroscopic fluid and the thin film in coexistence for the case of a first-order wetting transition, show that although the contact angle vanishes at the transition, the film thickness remains finite.

6. Dewetting of Solids

Consider a thin film of thickness D that uniformly covers a solid substrate except for a cylindrical hole of radius R where the solid is exposed to the air. What is the free energy of this film compared to the case of a perfect film without the hole in terms of D, R, and the relevant surface/interfacial tensions? What is the critical radius at which this hole will grow and why?

4.8   REFERENCES

1. See P. G. de Gennes, *Rev. Mod. Phys.* **57**, 827 (1985) and L. Leger and J. F. Joanny, *Rep. Prog. Phys.* **55**, 431 (1992) for reviews of both the statics and dynamics of wetting.

2. See M. Schick in *Liquids at Interfaces*, eds. J. Charvolin, J. F. Joanny, and J. Zinn-Justin, Les Houches Session XLVIII (North-Holland, Amsterdam and New York, 1990), p. 415 for a review with an emphasis on phase transitions and also the articles by A. M. Cazabat, ibid., p. 371, and D. Beysens, ibid., p. 499, for experimental discussions.

3. J. F. Joanny, *Physico-Chemical Hydrodynamics* **9**, 183 (1987).

4. T. Young, *Phil. Trans. R. Soc. London* **95**, 65 (1805).

5. F. Brochard, J. Di Meglio, D. Quere, and P. G. de Gennes, *Langmuir* **7**, 335 (1991).

6. J. F. Joanny and P. G. de Gennes, *J. Chem. Phys.* **81**, 552 (1984).

7. J. F. Joanny, *J. Theoretical and Appl. Mechanics* **23**, 249 (1986).

8. L. H. Tanner, *J. Phys. D* **12**, 1473 (1979).

9. F. Melo, J. F. Joanny, and S. Fauve, *Phys. Rev. Lett.* **63**, 1958 (1989).

10. H. E. Huppert, *Nature* **300**, 427 (1982).

11. A. M. Cazabat, F. Heslot, S. M. Troian, and P. Carles, *Nature* **346**, 824 (1990).

12. S. M. Troian, E. Herbolzheimer, S. A. Safran, and J. F. Joanny, *Europhys. Lett.* **10**, 25 (1989).

CHAPTER 5

Interactions of Rigid Interfaces

5.1 INTRODUCTION

The previous chapters presented a theoretical framework for the structure, fluctuations, and phase transitions of interfaces and surfaces, focusing on the properties of a single, isolated interface. Here, we consider the interactions between rigid (*i.e.,* flat) interfaces and surfaces. These interactions are a necessary first step in characterizing the properties of systems with a *bulk*, thermodynamic amount of interface such as colloidal dispersions and self-assembling structures. In addition, the interactions of only two interfaces can be studied with macroscopic force-balance experiments[1]. We begin with a brief review of the types of molecular interactions that are of interest in these systems and then focus on the types of long-range forces that are important in determining the large-scale properties of surfaces and interfaces. In our approach, the molecular, short-range interactions are taken into account via a generally phenomenological treatment of the *local* properties of the system (*e.g.,* surface or interfacial tension, bare correlation lengths, parameters in free energy expansions) while the long-range interactions, such as the van der Waals and electrostatic interactions, are considered explicitly since these interactions are fairly universal in nature and can be directly incorporated into the theory. We thus consider the microscopic origin of the van der Waals interactions between interfaces. We show how these interactions, which are *universal* in nature, lead to long-range forces and then discuss the continuum theory of these effects. Next, we treat electrostatic interactions between charged surfaces or interfaces with an emphasis on the connection between the free energy and the spatially dependent charge distribution. Finally, we consider the interaction between two surfaces that is mediated by a solute dissolved in a solvent that occupies the volume between the two surfaces; again, we shall see that this interaction is quite universal in nature. In this chapter, undulations or thermal fluctuations of the surfaces are neglected. The theory of interacting, fluctuating surfaces is treated in the following chapter on membranes. A comprehensive treatment of surface forces that also includes many aspects of experimental interest can be found in Ref. 1.

5.2 MOLECULAR INTERACTIONS

The following is a summary, based on Ref. 1, of the types of molecular interactions that are important in understanding the structure and phase behavior of surfaces and interfaces. The interactions in multicomponent systems with surfaces and interfaces are often related to the interactions between molecules in a particular type of medium. This is particularly important for self-assembling systems composed of surfactants or polymers, where the interactions and the subsequent equilibrium structures are strongly influenced by the type of solvent.

Strong Bonding: Covalent, Ionic, Metallic

Short-range interactions in uncharged systems are usually related to wave-function overlap (electron sharing) and are termed **covalent** bonds. The typical energies for a carbon-carbon double bond are $240\ k_BT$ and a single C-O bond is $136\ k_BT$. (Units are k_BT where $T = 300$K; note that 1 eV $\approx 40\ k_BT$.) In charged systems, *ionic* bonds arise from coulomb interactions (*e.g.*, Na^+Cl^-) and have a typical magnitude of $e^2/\epsilon a$ where e is the ionic charge and a is an interatomic spacing of several Å. In vacuum, where the dielectric constant, $\epsilon = 1$, this yields an energy of order $100\ k_BT$. In charged systems with delocalized electrons, *metallic* bonding is of significance. Typical cohesive energies are related to the width of the electron energy band and are of the order 1 or 2 eV ≈ 40 k_BT. Note that these large energies imply that such forces cannot be competitive with entropy at room temperature and that the melting involves the difference in free energy between the solid and liquid phases, not a total disruption of the molecular bonds.

Weak Bonding: Dipolar Interactions

In systems where the positive and negative charges are strongly bound on a single molecule (zwitterions), *dipolar* interactions between molecules can still occur. These interaction energies scale as $e^2 a^2 / r^3$ in a vacuum, where a is the spacing between the dipolar charges and r is the intermolecular distance. If the intermolecular distance is large, this yields energies of a few $k_B T$. This energy is an alignment energy that competes with the entropy of orientation. *Hydrogen bonding* is present in systems with OH, NH, FH, and ClH bonds. It is responsible for the relative stability of water, which is a low molecular weight molecule with a relatively high boiling point. Hydrogen bonding is found in systems that combine electronegative (negative charge) molecules or atoms and hydrogen (which has a positive polarization). Usually the hydrogen mediates interactions between molecules, with a typical strength of several $k_B T$ per bond.

Solvent-Mediated Energies

We now discuss the properties of molecules in a medium. For simplicity, we consider fluid media where the medium can be modeled as a continuum, dielectric. It is the polarizability of the solvent that promotes the solubility of molecules since charged molecules have a self-energy, related to the electrostatic energy of assembling the charges. This can be estimated in a simple classical model, by calculating the energy to charge a sphere of radius R (the ionic size) to a charge Z (the ionic charge). Since this energy scales like $(Ze)^2/\epsilon R$ we see that the energy is lowered if the molecule is placed in a medium with a higher dielectric constant. In a polar medium such as water, $\epsilon = 80$ and this energy is reduced by almost two orders of magnitude. This effect is partially responsible for the solubility of salt in water, but entropic interactions (free energy of $k_B T$ per degree of freedom) also enter. The solvation of polar molecules by the surrounding polar medium and the consequential lowering of the free energy of these polar molecules is known as the **hydrophilic** interaction. The interaction of polar molecules in a polar solvent is similarly affected by the polarizability of the solvent, which tends to reduce the direct interaction energy when compared with interactions in a vacuum.

Hydrophobic Interactions

Nonpolar molecules in polar solvents, such as water, have their free energy raised by the **hydrophobic** effect. The difference in van der Waals interactions is a minor (15% of free energy) effect compared with the fact that the nonpolar molecules restrict the entropy of the water; there are fewer ways for the water molecules to hydrogen bond in the presence of the nonpolar molecule. This entropic effect accounts for about 85% of the additional free energy to transfer a hydrocarbon to water. Since nonpolar molecules increase their free energy in polar solvents due to the disruption of the hydrogen bonding network of the water, these molecules tend to associate. An aggregate of such molecules disrupts the hydrogen bonding less than a solution where each molecule is individually dissolved. Such effects lead to the formation of **micelles** when surfactant molecules are dissolved in water. The polar part of such molecules (*e.g.*, $Na^+ SO_3^-$) are solvated in the water (hydrophilic) and the hydrocarbon parts are shielded from the water in the micelle interior. Since these interactions involve the local bonding of the solvent, they are short-range, in contrast to the electrostatic and dispersion (van der Waals) forces.

Van der Waals Interactions

Van der Waals or **dispersion** forces between molecules are important because they are universal — independent of the particular type of molecule; they are therefore present in all condensed matter systems. These forces arise from induced dipole-induced dipole interactions due to quantum mechanical fluctuations of the charge density. One can understand this interaction in a simple, but very approximate manner, if one considers two atoms that have spherically symmetric charge distributions and thus no average dipole moment. However, each atom does have an *instantaneous* dipole moment due to the charge separation of the positive nucleus and the valence electrons. This moment arises from quantum fluctuations in the valence charge density, which is only spherically symmetric on average. The electric field, $\vec{E}$, of this instantaneous dipole at a distance r from the atom, varies as $1/r^3$. Since the other atom is polarizable, it will respond to this field with an induced dipole moment, $\vec{p}$, which is proportional to this field and varies as α/r^3, where $\alpha > 0$ is the atomic polarizability. The interaction energy is proportional to $-\vec{p} \cdot \vec{E}$ and thus varies as $1/r^6$. The energy is attractive — *i.e.*, it decreases as r decreases — because the induced dipole tends to cancel the electric field that caused it.

The interactions are long range (power laws, varying as $1/r^6$) and are affected by the presence of the medium. The typical strength at a distance of 3 Å is 1 $k_B T$; they are thus, relatively weak interactions when compared with covalent or metallic bonding. On a molecular scale, these forces are responsible for the bonding of nonpolar solids such as rare gases (*e.g.*, argon) and hydrocarbons. However, due to their long range, these forces are of critical importance in understanding the free energies of composite materials and complex fluids where there may be domains composed of different components (*e.g.*, a layered dielectric). The equilibrium thickness of these regions as well as the interaction forces between these layers are determined by the long-range van der Waals interactions.

5.3 VAN DER WAALS INTERACTION ENERGIES

In this section we discuss the origin and implications of the dispersion interactions between molecules. We focus on the nonretarded interactions where the distance scale is small compared with the distance that light travels in a time associated with a typical atomic frequency — relativistic corrections are thus neglected. In addition, we consider only the energetics of the interactions; the thermal fluctuations of the electromagnetic field and the temperature dependence of these interactions are treated in the following section. After deriving the interaction between two atoms in a vacuum, we consider the *heuristic* extension of these concepts to interactions in a medium. Finally, we use these concepts to estimate the van der Waals interaction energies for macroscopic bodies. In Chapter 4, we saw how these long-range interactions, which can be either attractive or repulsive, can have important effects on the stability of thin films. In addition, dispersion forces determine the stability of colloidal systems as discussed in Chapter 7.

Quantum Mechanical Interactions

The van der Waals interactions between two bodies are essentially quantum mechanical in origin and can be understood by considering first the simple case of the interaction energy of two atoms[2] in a vacuum. We fix the centers of each atom at $\vec{r} = 0$ and $\vec{r} = \vec{R}$ respectively. The total Hamiltonian is

$$\mathcal{H} = \mathcal{H}_1 + \mathcal{H}_2 + V \tag{5.1}$$

where $\mathcal{H}_1$ and $\mathcal{H}_2$ are the Hamiltonians of the isolated atoms (they do not have to be identical in this treatment) and V is the interaction energy of the electrons on atom 1 with the positive nucleus of atom 2 and vice versa, as well as the interaction energies of the two electrons (we consider singly valent atoms for simplicity) and the two nuclei. The interaction of the electron of atom 1(2) with the nucleus of atom 1(2) is already included in $\mathcal{H}_{1(2)}$. As shown in Fig. 5.1, we label the two electron coordinates by $\vec{r}_1$ and $\vec{r}''_2$, respectively, we write:

$$V = e^2 \left[\frac{1}{|\vec{R}|} - \left(\frac{1}{|\vec{R} - \vec{r}_1|} + \frac{1}{|\vec{r}''_2|} \right) + \frac{1}{|\vec{r}_1 - \vec{r}''_2|} \right] \tag{5.2}$$

The quantum mechanical formula for the interaction energy, ΔU to second-order in perturbation theory is

$$\Delta U = \langle 0|V|0 \rangle + \sum_n \frac{|\langle 0|V|n \rangle|^2}{U_0 - U_n} \tag{5.3}$$

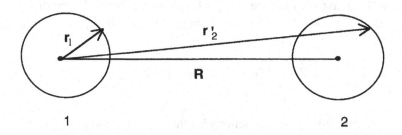

Figure 5.1 Coordinate system for calculating the van der Waals interaction between two atoms.

where the two-atom, unperturbed ground state ($|0\rangle$) energy is U_0 and the higher lying states ($|n\rangle$) have an energy U_n.

Nonretarded Interaction

When the distance between the two atoms is not too large ($R < c/\omega$, where ω is a typical atomic frequency), relativistic corrections do not have to be taken into account. One assumes that R is large enough, however, so that the overlap of the wavefunctions is small and one expands V in the small quantities, $\langle|\vec{r}_1|\rangle \ll R$ and $\langle|\vec{r}_2' - \vec{R}|\rangle \ll R$. We define $\vec{r}_2 = \vec{r}_2' - \vec{R}$ and find that the result of this expansion to lowest order is

$$V \approx -\frac{e^2}{R^3}\left[3(\vec{r}_1 \cdot \hat{R})(\vec{r}_2 \cdot \hat{R}) - \vec{r}_1 \cdot \vec{r}_2\right] \tag{5.4}$$

For spherically symmetric charge densities (*i.e.*, nonpolar atoms), the first-order term in the perturbation theory for ΔU, vanishes. The second-order term in Eq. (5.3) yields an attractive interaction that decays as $1/R^6$. The interaction between two atoms in a vacuum is always *attractive*, $\Delta U < 0$, since in Eq. (5.3) the ground state energy is less than the excited state energies. For large distances between the atoms (greater than 50 Å in vacuum), the van der Waals interaction requires relativistic corrections; it decreases faster than $1/R^6$ and is not always attractive. Similarly, in a dielectric medium, the interaction can be either attractive or repulsive.

An upper bound to (the negative quantity) ΔU can be found by noting that each term in Eq. (5.3) is negative and approximating the sum over excited states by the state with the energy closest to the ground state (*i.e.*, $n = 1$). We therefore write the energy of the first excited state as $U_1 = U_0 + \hbar\omega_1 + \hbar\omega_2$ where $\hbar\omega_1$ and $\hbar\omega_2$ are the energy differences for the two atoms respectively. In this approximation and for isotropic systems we find

$$\Delta U \approx \frac{-6\,e^4}{R^6}\,\frac{|\,\langle g_1|z_1|e_1\rangle\,\langle g_2|z_2|e_2\rangle\,|^2}{\hbar\omega_1 + \hbar\omega_2} \tag{5.5}$$

where $\langle g_i|$ and $\langle e_i|$ refer to the wavefunctions for the ith atom, and $i = 1, 2$, and $z_1 = \vec{r}_1 \cdot \hat{R}$, $z_2 = \vec{r}_2 \cdot \hat{R}$.

Interactions in a Medium

The matrix elements and energy denominators that enter into the van der Waals energy of Eq. (5.5) are related to the polarizability, α, which relates the dipole moment, $\vec{p} = e\vec{r}$ (with units of charge times distance), to an applied electric field[2]. For isotropic systems, the polarizability is a scalar quantity and we write:

$$\vec{p} = \alpha\vec{E} \tag{5.6}$$

The perturbation energy is $V = -\frac{1}{2}\vec{p}\cdot\vec{E} = -\frac{1}{2}\alpha E^2$ and one sees that the energy due to the electric field involves the same type of dipole matrix elements that enter into Eq. (5.5). In the approximation of a single excited state, one finds

$$\alpha_i = \frac{e^2|\,\langle g_i|z|e_i\rangle\,|^2}{\hbar\omega_i} \tag{5.7}$$

Thus, α has units of length cubed. We therefore can write the van der Waals energy in terms of the polarizabilities of each atom:

$$\Delta E \approx -\frac{6\beta_{12}\alpha_1\alpha_2}{R^6} \tag{5.8}$$

where $\beta_{12} = \hbar\omega_1\omega_2/(\omega_1 + \omega_2)$ is a factor coming from the difference in the energy denominators. We emphasize that this expression is derived for a simplified two-state model. However, the behavior of the van der Waals interaction with polarizability as calculated in more detailed treatments described below, is in semiquantitative agreement with the less rigorous treatment presented here.

We now present a heuristic extension of this relationship between the van der Waals energy and the atomic polarizability to the case of two atoms or molecules interacting in a dielectric medium. For the case just treated, namely a vacuum, the polarizability is related to the dielectric constant by $\alpha_i = (\epsilon_i - 1)/\rho_i$, where $1/\rho_i$ is the molecular volume of species i. In a medium of dielectric constant, ϵ_m, one can heuristically generalize this expression to write the *excess* polarizability

$\alpha_i = (\epsilon_i - \epsilon_m)/\rho_i$. It is therefore suggestive to rewrite Eq. (5.8) for the dispersion interaction, $U = \Delta E$, between two molecules separated by a distance $\vec{r}$ as

$$U \approx -\frac{1}{r^6} \frac{6\beta_{12}}{\rho_1\rho_2} \frac{(\epsilon_1 - \epsilon_m)(\epsilon_2 - \epsilon_m)}{\epsilon_m^2} \tag{5.9}$$

where the factor of ϵ_m^2 in the denominator comes from the fact that the electric field which polarizes each atom is reduced by the dielectric constant of the medium and the energy is proportional to the square of the field. We see from this expression that the *sign* of the dispersion interaction depends on the *differences* between the molecular and medium dielectric constants. Note that if $\epsilon_1 = \epsilon_2$ $U < 0$; two like molecules always attract. If $\epsilon_m > \epsilon_1$ and $\epsilon_m > \epsilon_2$ or if $\epsilon_m < \epsilon_1$ and $\epsilon_m < \epsilon_2$; the van der Waals interaction is also always attractive. Only in the intermediate case where ϵ_m lies between ϵ_1 and ϵ_2 can the van der Waals interaction be repulsive. It is traditional to define the Hamaker constant, A_{12}, for the interaction of two molecules in a given medium by the expression

$$U = -\frac{A_{12}}{\pi^2\rho_1\rho_2} \frac{1}{r^6} \tag{5.10}$$

where ρ_i is the density of molecule i. Typically the Hamaker constant, which has units of energy, is of order $25k_BT$. The discussion here suggests that the sign of A_{12} is related to the sign of the product $(\epsilon_1 - \epsilon_m)(\epsilon_2 - \epsilon_m)$ — e.g., for identical molecules, $A_{12} > 0$, and the interaction is attractive.

Interaction of Many Molecules

The slow nature of the decay of the dispersion interaction gives rise to a long-range interaction in large bodies. For a many-particle system the interaction energy can be obtained if one assumes that the van der Waals interactions are pairwise additive (see the following Section for a more rigorous treatment). We therefore consider the interaction, U, between two parts of the system with densities $n_1(\vec{r})$ and $n_2(\vec{r})$, given by

$$U = -W_0 \int d\vec{r}\, d\vec{r}'\, \frac{n_1(\vec{r})n_2(\vec{r}')}{|\vec{r} - \vec{r}'|^6} \tag{5.11}$$

where W_0 is related to the Hamaker constant. The long-range nature of these interactions can be seen by considering a single molecule above a surface at a distance D in the $\hat{z}$ direction. If the material with the surface has a density, n_0,

$$U = -W_0\, 2\pi n_0 \int_D^\infty dz \int_0^\infty d\rho\, \frac{\rho}{(z^2 + \rho^2)^3} \tag{5.12}$$

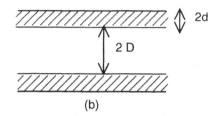

Figure 5.2 Van der Waals interactions across a vacuum and of two thin films.

Thus, the net interaction with all the surface decays slowly as $1/D^3$ and the force the molecule feels is attractive and decays as $1/D^4$.

Surface Interactions

In a similar manner, one can calculate the interaction energy per unit area, $u = U/A$, of two surfaces of area, A, separated (by a vacuum) over a distance $2D$ (see Fig. 5.2):

$$u = -W_0 (2\pi n_0^2) \int_D^\infty dz \int_{-\infty}^{-D} dz' \int_0^\infty d\rho \frac{\rho}{[(z-z')^2 + \rho^2]^3} \qquad (5.13)$$

One finds that $u = (-W_0 \pi n_0^2)/(48D^2)$. The interaction energy per unit area between two membranes of thickness, $2d$, separated by a distance, $2D$ is proportional to d^2/D^4 when $d \ll D$ and is attractive. For small separations relative to the membrane thickness, $D \ll d$, the interaction decays as $1/D^2$. In this case, the interaction decays with the same power law as in the case of two semi-infinite media separated by a gap of distance D, since when $d \gg D$, the membranes appear to be infinitely thick compared to the gap size. Another important quantity is the self energy of a slab of thickness D. For this calculation, a cutoff on the short distance part of the energy is needed, so one can write the dispersion interaction as: $[r^2 + a^2]^{-3}$, where a is the minimum distance between molecules. One finds in the energy per unit area a bulk term linear in D, a surface term, independent of D and a term that is proportional to $-A/D^2$, where A is the Hamaker constant. Thus a thin slab bounded on both sides by a vacuum, would tend to thin due to the van der Waals energy. If the slab is bounded on both sides by the same dielectric medium, the heuristic discussion presented previously indicates that the appropriate Hamaker constant scales as $(\epsilon_m - \epsilon)^2$, where ϵ_m is the dielectric constant of the film and ϵ is that of the surrounding medium; here too, the film would tend to thin.

Example: van der Waals Energy of a Thin Film

Compute the van der Waals energy of a thin film of thickness D, considering the bulk, surface, and interaction terms:

As before, we write the energy per unit area as

$$u = -2\pi n_0^2 W_0 \int_0^D dz \int_0^D dz' \int_0^\infty d\rho \; \frac{\rho}{[(z - z')^2 + \rho^2 + a^2]^3} \qquad (5.14)$$

The cutoff, a^2, in Eq. (5.14) is needed for this calculation since we do not want to count the infinite energy of a particle interacting with itself. Thus, a represents a typical, minimal intermolecular spacing. Performing the integral over ρ, we find

$$u = -\tfrac{1}{2}\pi n_0^2 W_0 \int_0^D dz \int_0^D dz' \frac{1}{[(z - z')^2 + a^2]^2} \qquad (5.15)$$

Performing the integration over z' and defining $\tilde{z} = D - z$ gives

$$u = -\frac{\pi n_0^2 W_0}{4a^2} \int_0^D dz \left[\frac{1}{a} \tan^{-1} \frac{\tilde{z}}{a} + \frac{\tilde{z}}{\tilde{z}^2 + a^2} + \frac{1}{a} \tan^{-1} \frac{z}{a} + \frac{z}{z^2 + a^2} \right] \qquad (5.16)$$

Finally, after integrating over z we find

$$u = -\frac{\pi n_0^2 W_0}{2a^3} D \; \tan^{-1} \frac{D}{a} \qquad (5.17)$$

Expanding for $x = D/a \gg 1$, where $\tan^{-1} x = \pi/2 - x^{-1} + x^{-3}/3$ for large values of x, we see that there is a bulk term, linear in the thickness, D, a surface term, independent of D, and an interaction term that is:

$$u_{int} = -\frac{\pi n_0^2 W_0}{6D^2} \qquad (5.18)$$

■

We now consider heuristically the more general case of a film of dielectric constant ϵ_m bounded by infinite half-spaces of dielectric constants ϵ_1 and ϵ_2, as shown in Fig. 5.3. A simple extension of the ideas presented above for the interaction of two molecules in a medium indicates that the Hamaker constant A is related to the product $(\epsilon_1 - \epsilon_m)(\epsilon_2 - \epsilon_m)$. There are several cases of interest:

(i) When $\epsilon_1 = \epsilon_2$, a thin film in a vacuum or between two identical semi-infinite media, $A > 0$ and the film tends to thin due to this interaction.

(ii) When $\epsilon_m > \epsilon_1$ and $\epsilon_m > \epsilon_2$ or if $\epsilon_m < \epsilon_1$ and $\epsilon_m < \epsilon_2$ — i.e., the two media are *both* either of higher or lower dielectric constant than the thin film $A > 0$ and the film again tends to thin due to this interaction; there is an effective attraction between the half-spaces. This is independent of the magnitudes of ϵ_1 and ϵ_2.

(iii) When $\epsilon_1 > \epsilon_m$ but $\epsilon_2 < \epsilon_m$, $A < 0$ and the film will tend to thicken; there is an effective repulsion between the half-spaces.

These considerations are important in determining the equilibrium thickness of a wetting layer. In the cases where the Hamaker constant is negative (self-attraction, compared with medium and substrate), this tends to thicken the wetting layer; in the opposite case, the wetting layer tends to thin. These conclusions can also be derived for the one-dimensional geometry discussed here by assuming that the van der Waals interactions of the different components are additive; the effective Hamaker constant will then be related to sums and differences for the Hamaker constants for the different media, in a manner similar to that described previously. In the following section, we will see how these qualitative ideas are confirmed by the detailed quantum electrodynamic treatment that views the van der Waals interactions as quantum-statistical fluctuations of the electromagnetic field in complex dielectric media.

5.4   CONTINUUM THEORY OF VAN DER WAALS FORCES

The theory for the van der Waals interactions presented in the previous section applies to macroscopic media only in a qualitative sense. This is because (i) the additivity of the interactions is assumed — i.e., the energies are written as sums of the separate interactions between every pair of molecules; (ii) the relationship of the Hamaker constant to the dielectric constant is based on a very oversimplified quantum-mechanical model of a two-level system; (iii) finite temperature effects on the interaction are not taken into account since it is a zero-temperature description. Here, we present a simplified derivation of the van der Waals interaction in continuous media, based upon arguments first presented[3] by Ninham *et al.* ; a more rigorous treatment can be found in Ref. 4. The van der Waals interactions arise from the free energy of the fluctuating electromagnetic

15. A treatment of self-assembly can be found in *Micelles, Microemulsions, Membranes and Monolayers*, eds. W. Gelbart, A. Ben-Shaul, and D. Roux (Springer-Verlag, New York, 1994).

16. For a survey of several areas in complex fluids see *Physics of Complex and Supermolecular Fluids*, eds. S. A. Safran and N. A. Clark (Wiley, New York, 1987), *Structure and Dynamics of Strongly Interacting Colloids and Supramolecular Aggregates in Solution*, eds. S. Chen, J. S. Huang, and P. Tartaglia (Kluwer, Boston, 1992).

17. D. Nelson, T. Piran, and S. Weinberg, *Statistical Mechanics of Membranes and Surfaces* (World Scientific, Teaneck, NJ, 1989).

18. J. N. Israelachvili, *Intermolecular and Surface Forces* (Academic Press, New York, 1992).

19. A. Halperin, M. Tirrell, and T. P. Lodge, *Adv. Polymer Sci.* **100**, 31 (1991).

20. D. Lasic, *American Scientist* **80**, 20 (1992).

21. C. A. Croxton, *Statistical Mechanics of the Liquid Surface* (Wiley, New York, 1980); J. P. Hansen and I. R. McDonald, *Theory of Simple Liquids* (Academic Press, New York, 1990).

22. D. Forster, *Hydrodynamic Fluctuations, Broken Symmetry and Correlation Functions* (Benjamin (Addison-Wesley), Reading, MA, 1975).

23. Light scattering in complex fluids is discussed in *Light Scattering in Liquids and Macromolecular Solutions*, eds. V. Degiorgio, M. Corti, and M. Giglio (Plenum, New York, 1980).

24. S. K. Ma, *Modern Theory of Critical Phenomena* (Benjamin (Addison-Wesley), Reading, MA, 1976); N. Goldenfeld, *Lectures on Phase Transitions and the Renormalization Group* (Addison-Wesley, Reading, MA, 1992).

25. P. J. Flory, *Principles of Polymer Chemistry* (Cornell University Press, Ithaca, New York, 1953).

26. F. M. Stein, *Introduction to Matrices and Determinants* (Wadsworth, Belmont, CA, 1967), ch. 6.

27. G. Arfken, *Mathematical Methods for Physicists* (Academic Press, New York, 1985).

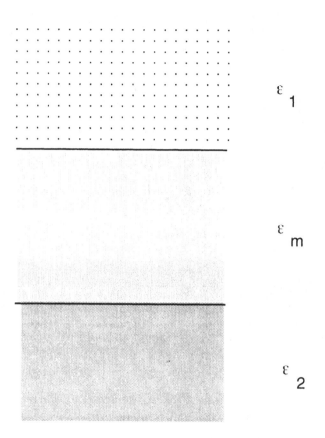

Figure 5.3 Van der Waals interactions of a thin film between two semi-infinite dielectric media.

field in the system. For bodies whose separations are much greater than molecular distances, we can use a continuum theory where the media are parameterized by their frequency dependent dielectric constants. Maxwell's equations are used to calculate the effect of the fluctuating field on distant points in space and hence the correlations between these fluctuations; this leads to a set of normal modes whose amplitudes are determined by the quantum Bose statistics of photons. Our goal is to consider the geometry shown in Fig. 5.3 for a thin film between two semi-infinite dielectrics and calculate the dependence of the free energy as a function of the film thickness. We shall see how the frequency dependence of the dielectric constant enters and changes the general thickness dependence from the simple $1/D^2$ law predicted earlier. In addition, this treatment justifies (under certain simplifying assumptions) the connection between the dielectric constants of the three media and the tendency for the van der Waals interaction energy to be repulsive or attractive — *i.e.,* tending to thicken or thin the intervening film. Readers interested only in the results can skip to the discussion of the special cases.

Free Energy

We consider at first the free energy of noninteracting photons in transparent media where the photons can be considered as an ideal gas of Bosons. The free energy of the electromagnetic normal modes in even a complex, inhomogeneous medium is expressed in terms of the Bose statistics[5] of the "ideal-gas" of photons. We consider normal modes of energy $\hbar\omega(\vec{q})$ where $\vec{q}$ is the wavevector, and $n_{\vec{q}}$ is the occupation number. The free energy is written in terms of the log of the partition function which is given by the sum over modes appropriate to the Bose gas:

$$F = -T \sum_{\vec{q}} \log \sum_{n_{\vec{q}}=0}^{\infty} e^{-\hbar\omega(\vec{q})(n_{\vec{q}}+1/2)/T} \tag{5.19}$$

Here, the factor of $1/2$ in the exponent accounts for the zero-point, quantum fluctuations of the system; the Bose statistics imply that any mode can have an occupation number that ranges from 0 to ∞. Performing the sum, we find

$$F = T \sum_{\vec{q}} \log \left[2 \sinh \frac{\hbar\omega(\vec{q})}{2T} \right] \tag{5.20}$$

For a layered system, where the inhomogeneity in the media and the dielectric constant is in the $\hat{z}$ direction only, we distinguish between the in-plane degrees of freedom and the perpendicular direction. Focusing on a system described by a length scale, L (*e.g.,* the thickness of a film bounded by two semi-infinite half-spaces of dielectric, as discussed before), the normal mode frequencies for a given branch, j, are functions of both $\vec{q}$ and L. We therefore take $\omega(\vec{q}) = \omega_L(\vec{q}; j)$ where $\vec{q}$ is a two-dimensional wavevector and the index j is the perpendicular degree of freedom. The index L indicates that we shall focus on those modes that are sensitive to the thickness of the film (*i.e.,* the surface modes). The free energy of interaction of the system per unit area, $\Delta f(L)$ (*i.e.,* the L dependent part of the free energy) can be written:

$$\Delta f(L) = \frac{1}{4\pi^2} \int \left[G_L(\vec{q}) - G_\infty(\vec{q}) \right] d\vec{q} \tag{5.21}$$

where the integration is over the two-dimensional wavevector, $\vec{q}$ and

$$G_L(\vec{q}) = T \sum_j \log \left[2 \sinh \frac{\hbar \omega_L(\vec{q};j)}{2T} \right] \tag{5.22}$$

The summation over j accounts for a multiplicity of normal modes, all having the same value of $\vec{q}$. To obtain the interaction energy, only those modes whose frequency explicitly depends on the film thickness, L, need be considered.

In the following example, we show that Eq. (5.22) can be written in the form

$$G_L(\vec{q}) = T \sum_{n=0}^{\infty}{}' \log D_L(\vec{q}, i\omega_n) \tag{5.23}$$

where

$$D_L(\vec{q}, \omega) = 0 \tag{5.24}$$

is the dispersion relation for the normal modes of interest. The prime indicates that the term with $n = 0$ is multiplied by a factor of 1/2 and $\omega_n = 2\pi n T/\hbar$, If we normalize the dispersion relation so that $D_{L \to \infty}(\vec{q}, \omega) = 1$ for all $\vec{q}$ and ω, then $G_\infty = 0$. The dispersion relation for the inhomogeneous system is obtained from Maxwell's equations[3,4] as discussed in the following section.

These results are used in Eq. (5.21) to derive the interaction free energy, which can be written in the general form

$$\Delta f(L) = \frac{T}{4\pi^2} \sum_{n=0}^{\infty}{}' \int \rho(\vec{q}) \, \log D_L(\vec{q}, i\omega_n) \, d\vec{q} \tag{5.25}$$

This form for the free energy is valid when the dispersion relation is written in a form where $D_{L \to \infty}(\vec{q}, i\omega_n) = 1$. The function $\rho(\vec{q})$ is the density of photon states in momentum space[4], which in our derivation for transparent media, can be taken as unity (using the definitions used in Ref. 4, $\beta = A/(4\pi^2)$, where A is the cross-sectional area).

Example: Relation of Free Energy to Dispersion Relation

Relate[3] Eq. (5.21) to Eq. (5.25) by using the relation between the normal mode frequency, $\hbar \omega_L(\vec{q};j)$ and the dispersion relation for these normal modes: $D_L(\vec{q}, \omega_L(\vec{q};j)) = 0$.

We use the identity for two analytic functions of ζ which is a complex variable:

$$\sum_j g(\zeta_j) = \frac{1}{2\pi i} \int_C g(\zeta) \frac{d \log D(\zeta)}{d\zeta} \, d\zeta \tag{5.26}$$

where C is the contour of integration that includes the zeros of the function D, denoted by ζ_j, and excludes the poles of $g(\zeta)$. We shall interpret ζ_j to be the normal mode frequency so that $D(\zeta_j) = 0$ is the dispersion relation for these modes. We now choose the function $g(\zeta)$ to be

$$g(\zeta) = \log \left[2 \sinh \frac{\hbar\zeta}{2T} \right] = \frac{\hbar\zeta}{2T} - \sum_{n=1}^{\infty} \frac{e^{-n\hbar\zeta/T}}{n} \tag{5.27}$$

where we have used a series expansion of the logarithm. Next, we perform the integration in Eq. (5.26) over a contour from $-i\infty$ to $i\infty$ along the imaginary axis and around the right half-plane along a semicircular path whose radius tends to infinity (excluding zeros of D on the imaginary axis).

Using the normalization that $D_L(\vec{q}, \omega \to \infty) = 1$ (since the free energy of Eq. (5.21) involves the log of the difference, this can be generalized as long as D tends to a constant as $\omega \to \infty$), we can write Eq. (5.22) using Eq. (5.26) as

$$G_L(\vec{q}) = \frac{T}{2\pi i} \int_{-\infty}^{\infty} g(i\zeta) \frac{d \log D_L(\vec{q}, i\zeta)}{d\zeta} \, d\zeta \tag{5.28}$$

and use Eq. (5.27) and an integration by parts to find

$$G_L(\vec{q}) = \tfrac{1}{2}\hbar \sum_j \omega_L(\vec{q}; j) + \frac{\hbar}{2\pi} \sum_{n=1}^{\infty} \int_{-\infty}^{\infty} \cos[n\hbar\zeta/T] \log D_L(\vec{q}, i\zeta) \, d\zeta + I_0 \tag{5.29}$$

where

$$I_0 = \frac{\hbar}{2\pi i} \sum_{n=1}^{\infty} \int_{-\infty}^{\infty} \sin[n\hbar\zeta/T] \log \left[D_L(\vec{q}, i\zeta) \right] \, d\zeta \tag{5.30}$$

As shown below, the dispersion relation is a function of the frequency dependent dielectric function, $\epsilon(\omega)$. When the dielectric function is an even function of ω, the dispersion relation is even as well, and $I_0 = 0$ by symmetry.

We thus evaluate Eq. (5.29) with $I_0 = 0$ by noting that the integral can be performed using the identity

$$\sum_{n=1}^{\infty} \cos nx = \pi \sum_{n=-\infty}^{\infty} \delta(x - 2\pi n) - 1/2 \qquad (5.31)$$

A further partial integration over the last term (with the factor $-1/2$) yields

$$G_L(\vec{q}) = T \sum_{n=0}^{\infty} {}' \log D_L(\vec{q}, i\omega_n) \qquad (5.32)$$

■

Electromagnetic Normal Modes

The evaluation of the interaction free energy of Eq. (5.25) requires the dispersion relation for the normal modes in an inhomogeneous medium. We consider the specific case, shown in Fig. 5.3, of a thin film of thickness L and of dielectric function $\epsilon_m(\omega)$ bounded by two half-spaces of dielectric functions $\epsilon_1(\omega)$ and $\epsilon_2(\omega)$. Following Refs. 3,4, we write the temporal Fourier transform of the electric field: $\vec{E}(\vec{r}, t) = \sum_{\omega} \vec{E}(\vec{r}, \omega)e^{-i\omega t}$, where the discrete sum is over the frequencies of the normal modes. A similar expression defines $\vec{H}(\vec{r}, \omega)$. From Maxwell's equations in a general dielectric medium, with uniform magnetic susceptibility and with no external charges, one can show that each Cartesian component of the electric or magnetic field satisfies the wave equation. Thus, we have, for example:

$$\nabla^2 \vec{E}(\vec{r}, \omega) + \frac{\epsilon\omega^2}{c^2} \vec{E}(\vec{r}, \omega) = 0 \qquad (5.33)$$

In addition, in the absence of external charges we have

$$\nabla \cdot \vec{E} = 0 \qquad (5.34a)$$

and

$$\nabla \cdot \vec{H} = 0 \qquad (5.34b)$$

The wave equation is solved subject to the boundary condition that the tangential (x and y) components of $\vec{E}$ and $\vec{H}$ are continuous and that the normal (z) components of $\epsilon\vec{E}$ and $\mu\vec{H}$ are continuous at the boundaries between the different regions. In each region, labeled by the subscript $j = 1, m, 2$, the solutions that are applicable to systems with translational symmetry in the $\hat{x}$ and $\hat{y}$ directions, have the form $f_j(z) \exp[i(q_x x + q_y y)]$ for each component of the fields; the modes are thus labeled by the wavevector $\vec{q} = (q_x, q_y)$. The wave equation dictates that the function $f_j(z)$ has the form

$$f_j(z) = A_j e^{\rho_i z} + B_j e^{-\rho_i z} \qquad (5.35)$$

where the integration constants A_j and B_j are found from the boundary conditions and we have defined

$$\rho_j(\omega)^2 = q^2 - \frac{\omega^2 \epsilon_j(\omega)}{c^2} \qquad (5.36)$$

Since in this one-dimensional case one has complete freedom to define the $\hat{x}$ and $\hat{y}$ directions in a convenient manner, it is simplest to consider the wavevector as lying in the $\hat{x}$ direction. One then has two cases: (i) where the $\vec{E}$ field is perpendicular to the plane formed by the two vectors $\vec{q}$ and $\hat{z}$ — i.e., $\vec{E}$ is in the $\hat{y}$ direction and (ii) where the magnetic field is perpendicular to this plane. In both of these cases, denoted as E-waves and H-waves respectively, both the wave equation and Eq. (5.34) are satisfied and the form of the solutions is simplified. One then constructs the solutions in all the regions and finds a system of equations of the form of Eq. (5.35). This system only has a solution when the determinant of the coefficients is equal to zero; this condition yields two dispersion relations[4], one associated with the E-waves and another with the H waves:

$$\Delta_E = 1 - R(\rho_1, \rho_m)R(\rho_2, \rho_m)e^{-2\rho_m L} \qquad (5.37a)$$

$$\Delta_H = 1 - R(\epsilon_m \rho_1, \epsilon_1 \rho_m)R(\epsilon_m \rho_2, \epsilon_2 \rho_m)e^{-2\rho_m L} \qquad (5.37b)$$

where

$$R(x, y) = \frac{(x - y)}{(x + y)} \qquad (5.38)$$

Thus, the function D_L which enters the free energy is defined by

$$D_L(\vec{q}, \omega) = \Delta_E \Delta_H \qquad (5.39)$$

This form obeys the normalizations mentioned previously: $D_{L \to \infty}(\vec{q}, \omega) = 1$ and $D_L(\vec{q}, \omega \to \infty) = 1$, so long as $\epsilon_j(\omega \to \infty) = 1$. The free energy of interaction is therefore given by Eq. (5.25) as

$$\Delta f(L) = \frac{T}{2\pi} \sum_{n=0}^{\infty} {}' \int_{0}^{\infty} dq \, q \, \left[\log \left[1 - R(\rho_1, \rho_m) R(\rho_2, \rho_m) e^{-2\rho_m L} \right] + \right.$$

$$\left. \log \left[1 - R(\epsilon_m \rho_1, \epsilon_1 \rho_m) R(\epsilon_m \rho_2, \epsilon_2 \rho_m) e^{-2\rho_m L} \right] \right] \tag{5.40}$$

where all of the functions R are evaluated at $\omega = i\omega_n = 2\pi i n T / \hbar$ and all of the ρ_j are functions of $\vec{q}$ from Eq. (5.36).

Van der Waals Interaction: Special Cases

We now consider Eq. (5.40) in some limiting cases in order to study the nature of the van der Waals interaction in the continuum, quantum statistical picture described here. We can already see from the form of Eq. (5.40) that the interaction will be dependent on the relations between the various dielectric constants and that the sign of the interaction can vary depending on this relationship.

1. *At high temperatures*, $T \gg \min[\hbar\omega_0, \hbar c/L]$, where ω_0 is a characteristic frequency at which the dielectric constants approach their high-frequency limit of unity, the fluctuations are classical. Since the contribution from the high-frequency region is zero ($R(\rho_i, \rho_j) = 0$ when $\rho_i - \rho_j = 0$), and because the $T/\hbar$ is much larger than the highest frequency where the dielectric constant differs from zero, only the $n = 0$ component in the sum is important; when T is very large all the dielectric functions are evaluated at $\omega_n = 0$, their static values. The integration in Eq. (5.40) is performed by defining $x = 2qL$ (when $\omega_n = 0, \rho_i = q$) and

$$\Delta f(L) = \frac{T}{16\pi L^2} \int_{0}^{\infty} dx \, x \log \left[1 - \frac{(\epsilon_1 - \epsilon_m)(\epsilon_2 - \epsilon_m)}{(\epsilon_1 + \epsilon_m)(\epsilon_2 + \epsilon_m)} e^{-x} \right] \tag{5.41}$$

where all the dielectric functions are evaluated at their static ($\omega = 0$) values. If the difference between the dielectric constants is not too large, the log can be expanded for small deviations from unity. One then performs the integral over x to find

$$\Delta f(L) = -\frac{T}{16\pi L^2} \left[\frac{(\epsilon_1 - \epsilon_m)(\epsilon_2 - \epsilon_m)}{(\epsilon_1 + \epsilon_m)(\epsilon_2 + \epsilon_m)} \right] \tag{5.42}$$

This expression has the same form as the heuristic approximation discussed before and the interaction decays as $1/L^2$. When $\epsilon_1 < \epsilon_m < \epsilon_2$ or $\epsilon_2 < \epsilon_m < \epsilon_1$ — i.e., when the dielectric constant of the film has a value in between those of the half-spaces, the interaction is repulsive and the film will tend to thicken. In all other cases, the interaction is attractive and the film will tend to thin. We also note that the interaction energy is proportional to T, which indicates that this is a purely *entropic* contribution coming from the classical part of the fluctuations of the electromagnetic field.

2. *At low temperatures*, where $T \ll \min[\hbar\omega_0, \hbar c/L]$, the variable $\omega_n = 2\pi n T/\hbar$ can be approximated by a continuum variable, ω, which varies from 0 to ∞ — *i.e.*,

$$\sum_{n=0}^{\infty} \to \frac{\hbar}{2\pi T} \int_0^{\infty} d\omega \tag{5.43}$$

When additionally, $L \ll 2\pi c/\omega_0$, where ω_0 is a characteristic frequency at which all the dielectric constants approach unity, $\rho_j \approx q$ and the first log term in Eq. (5.40) can be ignored. This approximation thus ignores retardation effects (see below) that arise because of the finite travel time of light across the film at the highest frequencies of interest. Integrating the second term by parts yields

$$\Delta f(L) = -\frac{\hbar}{32\pi^2 L^2} \int_0^{\infty} d\omega \int_0^{\infty} dx \, x^2 \left[\frac{(\epsilon_1 + \epsilon_m)(\epsilon_2 + \epsilon_m)}{(\epsilon_1 - \epsilon_m)(\epsilon_2 - \epsilon_m)} e^x - 1 \right]^{-1} \tag{5.44}$$

where $\epsilon_j = \epsilon_j(i\omega)$. Again, if the dielectric constants are not too different, the exponential term dominates and one can approximate this expression as

$$\Delta f(L) \approx -\frac{\hbar}{16\pi^2 L^2} \int_0^{\infty} d\omega \left[\frac{(\epsilon_1 - \epsilon_m)(\epsilon_2 - \epsilon_m)}{(\epsilon_1 + \epsilon_m)(\epsilon_2 + \epsilon_m)} \right] \tag{5.45}$$

The interaction depends on $\hbar$ and is independent of temperature — *i.e.*, it is due to the quantum mechanical fluctuations of the system; as above, the interaction energy falls off as $1/L^2$. Comparing the forms Eq. (5.42) and Eq. (5.45) for high and low temperatures, we see that the *sign* of the interaction is determined by the differences in dielectric functions. In the high-temperature classical limit, it depends primarily on the zero-frequency dielectric function, while in the low-temperature limit, it depends on an average of these differences over ω.

3. *When retardation effects* are important, one can obtain another useful approximation to $\Delta f(L)$. We split Eq. (5.40) into one part that explicitly takes into account the $n = 0$ modes, and another, that accounts for the sum $n = 1...\infty$. In the second term, we consider the case where even at the lowest frequency of interest, ω_1 (see Eq. (5.40) and subsequent definitions of ω), the dielectric constants of the three media are very similar. This applies either to the case where ρ_1, ρ_2, ρ_m are almost the same for all frequencies or to the case where the temperature is high enough so that ω_1 is such that all the materials are in

their high-frequency limits where the dielectric constants are almost unity. In this approximation one keeps the $n = 0$ terms exactly, but expands the arguments of R in Eq. (5.40) to lowest order in the differences in the dielectric constants. One can then write[6]:

$$\Delta f(L) \approx -\frac{T}{8\pi L^2} \left[\frac{1}{2} R_0(\epsilon_1) R_0(\epsilon_2) + \sum_{n=1}^{\infty} \int_{r_n}^{\infty} dx\, x\, e^{-x} R_n(\epsilon_1) R_n(\epsilon_2) \right]$$

(5.46)

where

$$r_n = 2\omega_n \sqrt{\epsilon_m} L/c \tag{5.47}$$

and

$$R_n(\epsilon) \approx \frac{\epsilon_m(i\omega_n) - \epsilon(i\omega_n)}{2\epsilon_m(i\omega_n)} \tag{5.48}$$

Performing the integral over the variable x, we find

$$\Delta f(L) \approx -\frac{T}{8\pi L^2} \left[\frac{1}{2} R_0(\epsilon_1) R_0(\epsilon_2) + \sum_{n=1}^{\infty} R_n(\epsilon_1) R_n(\epsilon_2)(1 + r_n) e^{-r_n} \right] \quad (5.49)$$

The first term in this expression is the zero-frequency attractive term, which decays like $1/L^2$. One sees that the importance of the second term depends on the value of r_n, which is a function of the thickness, L. If the dielectric functions are weak functions of frequency in the range where this second term contributes, one can regard the factors of R_n as constants and perform the geometric series. In the limit that $4\pi LT/(\hbar c) \ll 1$, one finds that this term scales like $1/L^3$; this is commonly referred to as the van der Waals interaction in the limit where the retardation effects dominate.

We note that in general, the van der Waals interaction can be attractive or repulsive, depending on the details of the dielectric functions. Moreover, retardation effects can introduce terms that scale with distance in ways (*e.g.,* $1/L^3$) other than the simple $1/L^2$ interaction for the case of a layer of thickness L in between two infinite, dielectric media. In particular, in the case where the $1/L^2$ term is attractive and the $1/L^3$ term due to retardation effects is repulsive, one can have a value of the film thickness L that minimizes the total van der Waals energy. The system, if unconstrained, will try, at least locally, to attain this optimal value of its thickness. Similar ideas have been used in discussions of surface melting[7], to estimate the equilibrium thickness of a thin layer of water in between bulk ice and bulk vapor at the triple point at which these three phases coexist; this layer is thought to be some tens of Angstroms thick, indicating incomplete surface melting.

5.5 ELECTROSTATIC INTERACTIONS

Charged Surfaces

Electrostatic interactions are important, and often dominant, in systems where there is some separation between charged groups and thus some finite-range, internal, electric field. For example, some surfactant molecules have a fixed charge, which remains attached to the molecule, and mobile counterions, which are solvated by polar solvents. In the $NaSO_3$ polar groups, the positively charged Na is mobile while the SO_3 group is the negative, fixed charge. The electrostatic interactions between two interfaces containing such groups are important in stabilizing colloidal particles, clays, micelles, and bilayers, which are often coated with surfactants or charged polymers. Usually these interactions are repulsive (except when the distances between the interfaces become of order of a molecular size) and hence prevent agglomeration when such "particles" are dispersed in solution. In describing these systems one speaks of the fixed charge at the surface of the colloidal particle or micelle and the delocalized or mobile counterion that lowers its free energy through the entropy it gains by being in solution. The problem of interest is to find the spatial distribution of counterions around the surface of fixed charges. The Coulomb attractions tend to *bind* the counterions close to this interface while their entropy of mixing with the solvent tends to distribute them in the solution. One finds that for highly charged interfaces, most of the counterions reside close to the fixed charges in a thin layer whose size is typically several Angstroms; the remaining counterions are in solution with a density that decreases slowly (*i.e.*, as a power law) as a function of the distance from the surface of fixed charge. It is these delocalized charges that give rise to long-range interactions of the charged interfaces. The fact that some of the charge is highly localized near the surfaces suggests the concept of an effective charge of the interface in discussing long-range interactions, which are important in determining the overall structure and phase behavior of the system.

In this section, we discuss the free energy of the system of fixed and mobile charges as a function of the local charge density and show how the minimization of this free energy determines the spatial dependence of the charge distribution of the mobile counterions. Thinking of the system in terms of the total free energy allows a wide range of problems to be treated. For simplicity, we focus on the simple case of fixed charges and mobile counterions in the limit where these counterions can be treated as a pointlike, dilute gas. In actual systems, deviations from this simple case can be important and effects such as excluded volume of the counterions (*i.e.*, finite size effects), the discrete nature of the surface charge distribution, and specific interactions between the counterions can be included by generalizing the free energy and performing the appropriate minimization. We consider the simplest case of fixed charges and counterions since the charge distributions show power-law, long-range, spatial decay. As we show below, this leads to long-range repulsions between two such surfaces of fixed charge and can be important in stabilizing colloidal suspensions.

However, colloidal systems often have a solvent such as water, with added salt which introduces additional mobile charges of both positive and negative signs (*e.g.,* Na^+Cl^-). Thus, in addition to the counterions, the salt ions participate in the screening of the fixed charge and one must consider three types of mobile charges: the counterions, the positive salt ions, and the negative salt ions — whose entropy will be distinct from that of the counterions if the salt and counterions are distinguishable molecules. The addition of salt (which experimentally is almost always present in water to some degree) allows the fixed charge to be more easily screened and changes the spatial dependence of the charge distribution from one that decreases as a power law to one that decreases exponentially as one moves away from the surfaces of fixed charge. The decay length of this exponential is known as the screening length and is inversely proportional to the square root of the equilibrium salt concentration in the bulk of the solution. A relatively small amount of added salt can have a drastic effect on the electrostatic repulsion of two charged surfaces; with no salt, the counterion screening is long-range while with added salt this screening and the resulting interaction between the surfaces become short-range. In a colloidal system, this reduction in the effective repulsion between charged colloidal surfaces by added salt can result in a destabilization of the system as discussed in Chapter 7.

The minimization of the free energy to determine the average charge distribution and the resulting effective interactions between the surfaces of fixed charge, presupposes that thermal fluctuations are unimportant and that the system is determined by the properties of the average, spatially varying, charge distribution. This is not always true and correlation effects in the distribution of mobile counterions due to deviations from the simple, mean-field distribution, can change the nature of the interactions between the surfaces of fixed charge, sometimes resulting in short-range *attractive* interactions[1].

Free Energy

Consider a system with a (mobile) counterion density given by $n(\vec{r})$ and a density of fixed ions given by $n_f(\vec{r})$. We shall explicitly take into account the opposite signs of the counterions and fixed charges, so we regard both n and n_f as positive quantities. If the counterions in the solution are treated as a dilute, ideal gas, the free energy of the system can be written:

$$F = T \int d\vec{r}\, n(\vec{r}) \left[\log n(\vec{r})v_0 - 1\right] + \frac{e^2}{2\epsilon} \int d\vec{r} d\vec{r}\,' \frac{n(\vec{r})n(\vec{r}\,')}{|\vec{r} - \vec{r}\,'|}$$

$$-\frac{e^2}{\epsilon} \int d\vec{r} d\vec{r}' \, \frac{n(\vec{r}) n_f(\vec{r}')}{|\vec{r} - \vec{r}'|} + \frac{e^2}{2\epsilon} \int d\vec{r} d\vec{r}' \, \frac{n_f(\vec{r}) n_f(\vec{r}')}{|\vec{r} - \vec{r}'|} \qquad (5.50)$$

Here, we assume that the solvent can be treated as a continuous, fluid, dielectric with dielectric constant ϵ. The first term in Eq. (5.50) is the entropy of mixing (v_0 is the volume per counterion), the second is the repulsion among the counterions, the third the attraction of the counterions to the fixed charges, and the last term is the repulsion of the fixed charges. The factors of 1/2 in front of the interactions between like charges are present to avoid double counting of those interactions. Charge conservation implies

$$\int n(\vec{r}) \, d\vec{r} = \int n_f(\vec{r}) \, d\vec{r} \qquad (5.51)$$

For convenience, we define the potentials:

$$\psi(\vec{r}) = \int \frac{n(\vec{r}')}{|\vec{r} - \vec{r}'|} d\vec{r}' \qquad (5.52a)$$

$$\psi_f(\vec{r}) = - \int \frac{n_f(\vec{r}')}{|\vec{r} - \vec{r}'|} d\vec{r}' \qquad (5.52b)$$

Note that the conventional electrostatic potential, $V(\vec{r})$ whose negative gradient is the electric field, $\vec{E} = -\nabla V$, is related to ψ by $V(\vec{r}) = e\psi/\epsilon$.

In the spirit of the local random mixing approximation, we neglect fluctuations and determine the counterion charge density through a functional minimization of the free energy, subject to the constraint of charge conservation. We thus minimize the grand potential, G:

$$G = F - T\mu \int n(\vec{r}) d\vec{r} \qquad (5.53)$$

where F is given by Eq. (5.50). We thus find

$$n(\vec{r}) = n_0 e^{-\ell \psi_t} \qquad (5.54)$$

where

$$n_0 = e^\mu / v_0 \qquad (5.55)$$

$$\psi_t = \psi + \psi_f \qquad (5.56)$$

and

$$\ell = \frac{e^2}{\epsilon T} \tag{5.57}$$

is known as the Bjerrum length; for a single charge in water, it is approximately 7 Å. Substituting this solution back into the free energy, Eq. (5.50), we find

$$F = \frac{T}{2} \int d\vec{r} \ \left[n(\vec{r}) + n_f(\vec{r}) \right] (\log n(\vec{r}) v_0 - 1) \tag{5.58}$$

Thus, the minimum free energy is essentially given by the half of the entropy of the counterions.

Counterion Distribution

To solve Eq. (5.54) for the spatial distribution of the counterions, we note that the potentials of Eq. (5.52) obey Laplace's equation:

$$\nabla^2 \psi = -4\pi n(\vec{r}) \tag{5.59a}$$

$$\nabla^2 \psi_f = 4\pi n_f(\vec{r}) \tag{5.59b}$$

Thus, defining $n_t = n - n_f$, we can write:

$$\nabla^2 \psi_t = -4\pi n_t \tag{5.60}$$

where n_f is the fixed charge distribution that is known from the structure of the interface (*e.g.*, colloidal particle, micelle, membrane) and $n(\vec{r})$ is related to the potential by Eq. (5.54). We thus have a nonlinear equation for the charge density or equivalently for the potential.

Charge Distribution Around a Single Interface

Consider a single, planar interface at position $z = 0$ with a uniform, fixed charge $n_f = \sigma_0 \delta(z)$. Defining

$$\phi = \ell \psi_t \tag{5.61}$$

we have from Eqs. (5.54,5.60):

$$\nabla^2 \phi = -4\pi \ell \left[n_0 e^{-\phi} - \sigma_0 \delta(z) \right] \tag{5.62}$$

The quantity n_0 is determined from the conservation law:

$$n_0 \int dz \, e^{-\phi} = \sigma_0 \qquad (5.63)$$

The solution of Eq. (5.62), which decays to zero as $z \to \infty$, is

$$\phi = 2 \log \left[(|z| + \lambda)\sqrt{2\pi n_0 \ell} \right] \qquad (5.64a)$$

$$n(z) = \frac{1}{2\pi \ell} \frac{1}{(|z| + \lambda)^2} \qquad (5.64b)$$

where the Gouy-Chapman length, λ, akin to a screening length, is given by:

$$\lambda = \frac{1}{\pi \ell \sigma_0} \qquad (5.64c)$$

The charge density falls off slowly at large distances; however, most of the charge is localized in a layer of width λ, which tends to zero as the fixed-charge density, σ_0, goes to infinity.

Interaction of Two Charged Surfaces

More insight can be obtained by considering two charged surfaces as shown in Fig. 5.4. We now take $z = 0$ to lie in the midplane of the two surfaces which are separated by a distance $2D$. In the region between the surfaces, Eq. (5.62) is still obeyed, but the boundary condition is now that the charge distribution and the potential must be symmetric at $z = 0$ (equivalent to zero electric field on that plane). The potential that satisfies

$$\nabla^2 \phi = -4\pi n_0 \ell e^{-\phi} \qquad (5.65)$$

in the region $-D < z < D$ and obeys the boundary condition $d\phi/dz = 0$ at $z = 0$ is

$$\phi = \log \cos^2 k_0 z \qquad (5.66)$$

where

$$k_0^2 = 2\pi n_0 \ell \qquad (5.67)$$

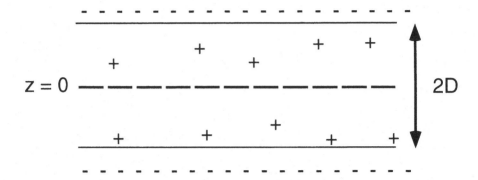

Figure 5.4 Two sheets with fixed negative charges and positive counterions. The dashed line represents the plane $z = 0$ and the distance between the sheets is $2D$.

The charge density is now

$$n(z) = \frac{n_0}{\cos^2 k_0 z} \tag{5.68}$$

where n_0 is the charge on the plane $z = 0$. The density n_0 is calculated from the conservation law and is easily obtained if one now considers a periodic array of such planes so that

$$\int_{-D}^{D} n(z)\, dz = \sigma_0 \tag{5.69}$$

This implies that

$$n_0 = \frac{k_0 \sigma_0}{2 \tan k_0 D} \tag{5.70}$$

Eqs. (5.67,5.70) imply

$$k_0 D \tan k_0 D = \pi \ell \sigma_0 D \tag{5.71}$$

this expression can be used to solve for the parameters k_0 and hence n_0 in terms of the charge density, σ_0, and the interlayer spacing, D. We now consider the two limits of small and large charge densities.

1. *Ideal Gas Limit:* In the limit of low charge density, $D\ell\sigma_0 \ll 1$, Eqs. (5.67,5.70) imply that $k_0 D \ll 1$, so that $n \approx \sigma_0/2D$ and is only weakly dependent on z.

This is the **ideal gas limit** where the counterions are nearly uniformly distributed between the two layers.

2. *High Charge Density Limit:* In the limit of high charge density, $D\ell\sigma_0 \gg 1$ (large separations, high charge densities), $k_0 D \approx \pi/2$ and Eq. (5.67) implies that $n_0\ell \sim 1/D^2$. For a uniform charge distribution, $n_0 \approx \sigma_0/2D$ so this result tells us that the charge at the midplane, $z = 0$, is smaller than that of a uniform distribution by an additional factor of $1/(\ell\sigma_0 D)$. As the spacing between the interfaces is made larger, the *ratio* of the charge at the midplane to a uniform charge distribution decreases. This result is strongly geometry dependent. These results suggest that one can think of an effective "fixed" charge, Z^*, given by the actual fixed charge ($Z = \sigma_0 A$, where A is the area of the sheet) minus the charge of those counterions localized near the fixed charges. For distances much greater than λ, the system approximately behaves as if the effective mobile charge

$$Z^* \approx Z\left(\frac{1}{\ell\sigma_0 D}\right) \tag{5.72}$$

of counterions were uniformly distributed in the space between the effective fixed charges. Thus, the long-distance properties determined by these interactions are strongly influenced by the reduction of the coulomb interactions by these effects.

Debye-Hückel Approximation

We now consider the charge distribution between two charged plates with (*e.g.*, a negative) fixed charge for the case where there is additional salt in the solution. The salt is assumed to be in equilibrium with a bulk salt solution of concentration (number of charges per unit volume), n_s. All of the charges in solution, including the counterions, positive salt ions (which we assume to be the same chemical species as the counterions), and negative salt ions participate in the screening of the fixed charges on the plates and our goal is to find the self-consistent charge distribution. Because there are more ions available to screen the fixed charges, we expect that the charge distribution will not be as long-range as that of the system with no salt, discussed earlier. In particular, we find below that the charge distribution decays exponentially with the distance between the plates; the length that characterizes this decay is known as the screening length and its inverse, κ, is related to the bulk salt concentration by the relationship:

$$\kappa^2 = 8\pi n_s \ell \tag{5.73}$$

The indistinguishability of the positive salt ions and the counterions means that one can consider both of them as comprising the same ideal gas. We construct

the free energy similar to that of Eqs. (5.50,5.53) including the entropies of all
the ions as well as the chemical potential term of the form

$$T\mu \int d\vec{r}\,(n_+ + n_-) \tag{5.74}$$

where n_+ and n_- are the positive and negative charge densities of the salt ions.
This term assigns the same value of the chemical potential to the positive and the
negative ions since the total number of each ion is the same by charge neutrality.
For a system in equilibrium with a *bulk* salt solution, the counterions (whose
number scales with the cross-sectional area, A) arising from the fixed charges
are negligible compared with the total number of salt ions (whose number scales
with the total volume of the bulk solution). We therefore regard n_+ as including
all the positive charges at this point. In addition to this conservation law, which
accounts for the chemical equilibrium of the salt ions, we must have charge
neutrality in the region between the two plates that have a surface charge density
of σ, implying:

$$\int_{-D}^{D} (n_+ - n_-)\,dz = \sigma \tag{5.75}$$

Minimizing the free energy as before, we obtain a Poisson-Boltzmann equation where the net charge

$$\rho(\vec{r}) = n_+ - n_- \tag{5.76}$$

arises from the local difference between the positive and negative ions only (the
minus sign explicitly takes into account the sign of the charge):

$$\nabla^2\phi = -4\pi\ell\rho(\vec{r}) = -4\pi\ell n_s(e^{-\phi} - e^{\phi}) \tag{5.77}$$

Where $\ell = e^2/(\epsilon T)$ is the Bjerrum length, $\phi = \ell\psi(\vec{r})$ is the dimensionless
potential with

$$\psi = \int d\vec{r}'\,\frac{\rho(\vec{r}')}{|r - r'|} \tag{5.78}$$

In Eq. (5.77) we have set the potential at $z \to \pm\infty$ to zero, so that the charge
density, which obeys $n_+ = n_0 \exp(-\phi)$ (with n_0 a constant that is determined by
the chemical potential) becomes $n_+ \to n_s$ as $z \to \pm\infty$ where n_s is the charge
density of the bulk. This fixes $n_s = n_0$.

   To solve for the potential, we consider the limit of high salt concentration
so that the fixed charges are well screened. We assume that the potential is small
(more precisely that it does not change much from the plates to the center) and

linearize Eq. (5.77). This approximation is called the Debye-Hückel approximation. We then find

$$\nabla^2 \phi = 8\pi \ell n_s \phi = \kappa^2 \phi \tag{5.79}$$

This is solved using the symmetry boundary condition that $\partial\phi/\partial z = 0$ at the
midplane, $z = 0$, and Eq. (5.75) to yield

$$\phi = -\frac{\sigma\kappa}{4n_s} \frac{\cosh(\kappa z)}{\sinh(\kappa D)} \tag{5.80}$$

and

$$\rho(z) = \tfrac{1}{2}\sigma\kappa \frac{\cosh(\kappa z)}{\sinh(\kappa D)} \tag{5.81}$$

For large values of κD the charge density is smaller at the midplane, $z = 0$,
compared to its value on the plates, at $z = \pm D$, by a factor proportional to
$\exp(-\kappa D)$; this strong screening, compared to the weaker, power-law behavior
found in the no-salt case, is due to the participation of the salt ions in the screening
of the fixed charge.

Forces between surfaces

One often investigates the nature of surfaces and interfaces by measuring the
forces of interaction between two macroscopic surfaces[1]. For charged systems,
these surfaces may be coated with a charged surfactant or polymer and they are
separated by a polar solvent in which the counterions are dissolved. It is thus of
interest to calculate the force between two surfaces. These forces are defined via
the changes in free energy of the system as the volume of the system is varied,
but the number of particles (*e.g.*, fixed charges and counterions) is kept constant.

The force in the $\hat{z}$ direction per unit area of surface, located at $z = D$,
whose normal is in the $\hat{z}$ direction is the **longitudinal pressure**, $\Pi_\ell(z = D)$,
which is related to the derivative of the free energy with respect to the changes
in the volume due to an expansion or compression in the $\hat{z}$ direction. Since
the system is anisotropic, one must differentiate between volume changes in
the longitudinal ($\hat{z}$) and transverse ($\hat{x}$, $\hat{y}$) directions. The force per unit area is
known as the stress, and for anisotropic systems the stress is a tensor quantity
whose components are denoted by $\Pi_{ij}(\vec{r})$. The first index denotes the direction
of the force and the second, the direction of the normal to the surface upon
which this force acts. Since systems can have shear restoring forces, there can
be off-diagonal components to the tensor Π. The divergence theorem tells us[8]
that the total force, $\vec{T}$, acting on a volume V, can be described by a volume

integral ($\int dV$) of the divergence of the stress or by a surface integral ($\int dA$) of the stress. In component notation,

$$T_i = \int dV \frac{\partial \Pi_{ij}}{\partial r_j} = \int dA \, \Pi_{ij} \, n_j \tag{5.82}$$

where $\vec{r} = (x, y, z)$ and where we use the convention that one sums over the repeated index, j. The normal to the surface in the jth direction is denoted by n_j. The total work performed by displacing the surface by a distance $\Delta \vec{R}$, is given by $\vec{T} \cdot \Delta \vec{R}$.

In our case of two infinite surfaces which are separated by a medium that transmits stress (*e.g.*, the charged fluid discussed earlier), the force per unit area in the $\hat{z}$ direction between two rigid surfaces of area A whose normals are in the $\hat{z}$ and $-\hat{z}$ directions, is therefore given by $\Pi_{zz}(z = D) = \Pi_\ell(D)$ where the subscript ℓ denotes that this is the longitudinal component of the stress tensor. The transverse component of the stress tensor, $\Pi_t(z) = \Pi_{xx}(z) = \Pi_{yy}(z)$ (for a system that is isotropic in the plane perpendicular to the $\hat{z}$ direction), is related to the force in the $\hat{x}, \hat{y}$ direction for displacements that increase the cross-sectional area of the system. The relevant normal vectors are also in the $\hat{x}, \hat{y}$ directions since one imagines increasing the cross-sectional area by exerting forces on the "sidewalls" containing the system.

One can calculate the longitudinal stress either by calculating $\Pi_\ell(z)$ at any point z and evaluating it for $z = D$, or by noting that the work per unit area done in displacing each surface by a distance (in the $\pm\hat{z}$ direction, respectively) ΔD, is given by $[\Pi_\ell(D) + \Pi_\ell(-D)] \Delta D$. This is equal to the change in free energy; noting that $\Pi_\ell(D) = \Pi_\ell(-D)$ by symmetry, we can therefore also calculate the force between the surfaces by calculating

$$\Pi_{zz}(D) = -\tfrac{1}{2} \frac{\partial f_s}{\partial D} \tag{5.83}$$

where f_s is the free energy per unit area. When $\Pi_\ell(D) > 0$, the free energy decreases as the volume increases and the force between the walls is repulsive and when $\Pi_\ell(D) < 0$, the force between the walls is attractive.

The transverse stress, Π_t, is calculated by finding the change in the total free energy as the volume is changed by increasing or decreasing the cross-sectional area. For a system described by a density, $n(z)$, which varies only in the $\hat{z}$ direction and is uniform and isotropic in the $\hat{x}, \hat{y}$ directions, we write the total free energy as

$$F = \int dz \, dA \, f[n(z)] \tag{5.84}$$

where the area element is dA and the free energy per unit volume, $f[n(z)]$ depends only on $n(z)$. When the cross sectional area is varied, but the total number of particles is kept constant, the total free energy changes in several ways: (i) the area element dA is changed, (ii) the *local* density changes since the density is inversely proportional to the local area element, and (iii) the boundary conditions may change as the area element is varied. All these changes are accounted for in the calculation of the transverse pressure of a gas of counterions discussed below.

Longitudinal Stress of Two Charged Surfaces

Using these ideas, we can calculate the longitudinal stress, $\Pi_\ell = \Pi_{zz}(D)$, arising from the force per unit area of two charged plates from Eq. (5.83). We consider the one-dimensional problem with cross-sectional area A and write the free energy per unit area f_s (see Eq. (5.50)). The longitudinal stress is then $\Pi_{zz}(D) = -\frac{1}{2}(\partial f_s/\partial D)$ where

$$f_s = T \int_{-D}^{D} dz \left[n(z) \left[\ln v_0 n(z) - 1 \right] + \frac{\ell}{2} n_t(z) \psi_t(z) \right] \tag{5.85}$$

where here $n(z)$ is the counterion charge density, but n_t and ψ_t refer to the total charge and the potential defined after Eq. (5.59). Since charge conservation implies that $\int n_t(z)\, dz = 0$, we see that adding any constant to the total potential does not change the free energy; we thus choose the total potential, $\psi_t(D) = \psi_t(-D) = 0$ on the surfaces of fixed charge. We next rescale the problem so that the distance is dimensionless and take the derivatives in Eq. (5.83) explicitly with respect to D. For this to work, the boundary conditions must be independent of D; this is not true for the condition of conservation of charge. We note, however, that if we transform the independent variable from the charge density, n, to the electric field, the boundary conditions satisfied by the field are independent of D. We therefore use the relationship between the potential, ψ_t, and a rescaled electric field, $\vec{\varepsilon}$ (not to be confused with the dielectric constant, ϵ), defined by $\vec{\varepsilon} = -\nabla \psi_t$ and

$$\frac{\partial^2 \psi_t}{\partial z^2} = -\varepsilon_z = -4\pi n_t(z) \tag{5.86}$$

to find that

$$n_t = \frac{1}{4\pi}\varepsilon_z \tag{5.87}$$

In the region just outside the charged surfaces, we have $n_t = n(z)$, the density of counterions. In this region, the conservation of charge, $\int n(z)\, dz = \sigma$, becomes a boundary condition on ε:

$$\varepsilon(D) - \varepsilon(-D) = 4\pi\sigma \tag{5.88}$$

(where by $\varepsilon(D)$ we mean the limit as $z \to D$ so as not to include the discontinuities in the field due to the fixed charges). For a symmetric problem, $\varepsilon(D) = -\varepsilon(-D)$ implying that

$$\varepsilon(z = 0) = 0 \tag{5.89}$$

These two boundary conditions are invariant to a rescaling of the z coordinate.

We first perform a partial integration on the second term in Eq. (5.85) to write:

$$\int dz\, n(z)\psi(z) = -\int dz\, \frac{1}{4\pi}\psi_{zz}\,\psi = \int dz\, \frac{1}{4\pi}(\psi_z)^2 = \int dz\, \frac{1}{4\pi}\varepsilon^2 \tag{5.90}$$

where we have omitted the subscript t in both n and ψ for clarity and have used the condition that the potential vanishes at the boundaries, $z = \pm D$. We thus find

$$f_s = T\int_{-D}^{D} dz\, \left[n(z)\,[\ln v_0 n(z) - 1] + \frac{\ell}{8\pi}\varepsilon^2 \right] \tag{5.91}$$

Next, we rescale the spatial variable so that $w = z/D$ and find

$$f_s = \frac{T}{4\pi} \int_{-1}^{1} dw\, \left[\varepsilon_w \left(\ln \frac{\varepsilon_w v_0}{4\pi D} - 1 \right) + \frac{\ell D}{2}\varepsilon^2 \right] \tag{5.92}$$

From Eq. (5.83) we find that the pressure is given by

$$\Pi_{zz}(D) = \frac{T}{8\pi} \int_{-1}^{1} dw\, \left[\frac{\varepsilon_w}{D} - \frac{\ell}{2}\varepsilon^2 \right] \tag{5.93}$$

But, the Euler-Lagrange equation, corresponding to the minimization of f_s with respect to ε yields

$$\varepsilon_{ww} - \ell D\,\varepsilon\,\varepsilon_w = 0 \tag{5.94}$$

This equation can be integrated once and the boundary condition that $\varepsilon = 0$ at $w = 0$, can be used to find

$$\frac{\varepsilon_w}{D} - \frac{\ell\varepsilon^2}{2} = \frac{\varepsilon_w(0)}{D} = \varepsilon_z(0) = 4\pi n(0) \tag{5.95}$$

where $n(0)$ is that value of the charge density at the midplane. Thus Eq. (5.93) becomes

$$\Pi_{zz}(D) = Tn(0) \tag{5.96}$$

and the longitudinal stress is given by the ideal-gas pressure of the charge density at the midplane, where the pressure due to the electric field vanishes. In the low charge density limit discussed earlier, this pressure is indeed the pressure of the ideal gas, $\Pi_{zz}(D) = T(\sigma/2D)$, confined between the layers. In the high charge density limit, $n_0 \sim 1/(\ell D^2)$ and the pressure acting on the confining layers is greatly reduced compared to the ideal gas limit. Equation (5.96) shows that this is due to the sharp reduction in the charge density far from the fixed charge planes, in the high charge density limit.

Transverse Stress of Two Charged Surfaces

The transverse stress or pressure, $\Pi_t(z) = \Pi_{xx}(z)$ is just related to the change in the total free energy as the volume is changed via a change in cross-sectional area only — i.e., the distance between the plates is kept constant and the number of particles is kept constant. Consider the expression for the free energy per unit area given by Eq. (5.91) with the boundary conditions, Eqs. (5.88,5.89). The boundary condition Eq. (5.88) depends on the charge per unit area, σ, and hence on the cross-sectional area (since the total charge is held fixed). To calculate the change in free energy as the cross-sectional area is changed, we must rescale the problem so that the boundary conditions are invariant with respect to changes in the cross-sectional area. We define a new independent variable, $\eta = \varepsilon/\sigma$, where ε is the rescaled electric field defined by $\vec{\varepsilon} = -\nabla\psi_t$. The boundary conditions are now independent of the cross-sectional area and we can calculate the pressure by taking the appropriate derivatives of the free energy. In terms of the variable η, the boundary conditions now read:

$$\eta(D) - \eta(-D) = 4\pi \tag{5.97}$$

$$\eta(z = 0) = 0 \tag{5.98}$$

From Eqs. (5.87,5.91) we can write the free energy per unit length, f_ℓ, in terms of η as

$$f_\ell = A\frac{T}{4\pi}\left[\sigma\eta_z\left(\ln\frac{\eta_z\sigma v_0}{4\pi} - 1\right) + \frac{\ell}{2}\sigma^2\eta^2\right] \quad (5.99a)$$

where A is the cross-sectional area. Note that we have now written the free energy in unscaled units (*i.e.*, the coordinate in the longitudinal direction is $z = wD$, since the variation in volume does not involve the spacing D). Writing $\sigma = N_f/A$, where N_f is the number of fixed charges, we exhibit explicitly the dependence of the free energy on the cross-sectional area; our rescaled boundary conditions imply that η does not vary when the cross-sectional area is changed.

$$f_\ell = \frac{TN_f}{4\pi}\left[\eta_z\left(\ln\frac{\eta_z N_f v_0}{4\pi A} - 1\right) + \frac{\ell N_f}{2A}\eta^2\right] \quad (5.99b)$$

The transverse pressure is given by

$$\Pi_t = -\frac{\partial f_\ell}{\partial A} \quad (5.100)$$

After taking the derivative with respect to A, we write Π_t in terms of the scaled electric field, $\varepsilon = \eta\sigma$:

$$\Pi_t = \frac{T}{4\pi}\left[\varepsilon_z + \frac{\ell}{2}\varepsilon^2\right] \quad (5.101)$$

Using Eqs. (5.87,5.95) we write Π_t in terms of the charge density and find

$$\Pi_t = T\left(2n(z) - n(0)\right) \quad (5.102)$$

We note that while the longitudinal pressure is a constant, independent of z, as required by mechanical equilibrium, the transverse pressure does depend on z. In the limiting case of an isotropic system (*i.e.*, with *no* dependence of $n(z)$ on z), $\Pi_t = \Pi_\ell = Tn(0)$; we recover the ideal gas law.

5.6   SOLUTE-INDUCED INTERACTIONS

The discussion of the electrostatic interactions of two charged surfaces with an intervening solution of counterions is a specific case of the more general problem of solute-induced interactions. Namely, one considers a solute in a solvent that is bounded by two surfaces. The interactions of the solute with the walls, solvent, and with each other determine the free energy of the system of solvent, solute, and walls. From the dependence of the free energy on the distance between the walls we can obtain the sign and magnitude of the forces between the two walls. The electrostatic case focuses on long-range interactions. In this section, we show that even solutes that have short-range interactions induce an effective attractive interaction between the bounding surfaces. The interaction is attractive irrespective of whether the walls attract or repel the solute since the origin of the attraction is the composition gradient produced by the presence of the walls. (When the walls repel the solute, this is called a **depletion interaction** induced by the solute.) The overall gradient energy is reduced if the distance between the two walls is reduced and the attraction is generated because the osmotic pressure in the region between the two walls is lower than in the bulk region outside the walls; this difference in pressures results in a force that makes the two walls come together.

Model Free Energy and Boundary Conditions

We consider a one-dimensional model system where two walls at $z = \pm D$ bound a section of a binary solution composed of solute molecules in a solvent. We assume that the walls are permeable to the solute molecules so that the solution in the region between the walls is in equilibrium with the bulk. The concentration of solute in the bulk solution is denoted by c_b. The presence of the walls at $z = \pm D$, changes the local composition of the system, $c(z)$, which is assumed to vary only in the $\hat{z}$ direction. The bulk free energy per unit area, in units of $k_B T$, is a function of the local composition, $c(z)$ of the solute. We consider the change in the system free energy due to the changes in the concentration profile due to the interactions of the solute with the walls. Following the development in Chapters 1 and 2, we write the free energy difference, Δf (per unit area) between the bulk free energy of the system with the walls, and the free energy of the bulk solution (with no walls) in the region $-D < z < D$, as

$$\Delta f = \int_{-D}^{D} dz \left[f_b[c(z)] - f_b[c_b] + \tfrac{1}{2} B \left(\frac{\partial c(z)}{\partial z} \right)^2 \right] \tag{5.103}$$

where $f_b[c(z)]$ is the bulk free energy which is a functional of the local density, and $c = c_b$ is the bulk value of the concentration of solute. The total free energy

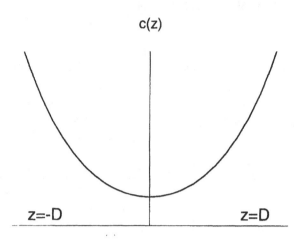

Figure 5.5 Concentration profile, $c(z)$, for the case where the solute is attracted to the walls. The walls are located at $z = \pm D$.

differs from Δf by a term involving the free energy density of the bulk solution integrated over the entire system; this term is independent of D and hence is dropped. In Eq. (5.103), $f_b[c(z)]$ is a functional of the local composition and the term proportional to B accounts for the short-range interaction energies that arise when there are composition gradients. In addition to this free energy, there is a contribution that comes from the interaction of the solute with the walls. We write this contribution to the free energy per unit area as

$$f_0 = f_w[c(D)] + f_w[c(-D)] \qquad (5.104)$$

For the symmetric problem treated here, $c(D) = c(-D)$. The wall-solute interaction energy, f_w is assumed to depend only on the *local* value of the composition of the solute molecules at the walls due to the short-range nature of the solute-wall interaction. The details of the function $f_w[c]$ depend on the specific interactions with the walls. For example, $f_w = -a_1 c$ would represent an attraction of the solute to the walls for $a_1 > 0$ and a repulsion for $a_1 < 0$; the attraction would just be proportional to the number of molecules adsorbed. The total free energy per unit area is the sum of $\Delta f + f_0$.

We proceed as we did in the discussion of electrostatic forces and consider the nondimensional variable $w = z/D$. We shall take all derivatives with the constraint that the system remain in chemical equilibrium. We thus add a chemical potential term to both the free energy of the inner region and the bulk. Defining $\Pi_b[c] = \mu c(z) - f_b[c]$ as the local osmotic pressure and μ as the chemical potential, we write the total grand potential per unit area, g, as

$$g = 2f_w[c(w = 1)] + D\left[\int_{-1}^{1} dw\left[-\Pi_b[c(w)] + \Pi_b[c_b] + \frac{B}{2D^2}\left(\frac{\partial c(w)}{\partial w}\right)^2\right]\right]$$

(5.105)

The condition of chemical equilibrium requires that the chemical potential is the same throughout the system and is thus given by its value for the bulk solution, where the composition is c_b:

$$\mu = \left(\frac{\partial f_b[c]}{\partial c}\right)_{c_b}$$

(5.106)

This fixes μ in terms of c_b. The boundary conditions on the minimization of g with respect to the functional $c(w)$ are (i) the symmetry condition, (see Fig. 5.5), $\partial c/\partial w = 0$ at the midplane at $w = 0$ and (ii) $\partial g/\partial c(w = 1) = 0$, the minimization of the grand potential with respect to the value of c on the wall. We regard $c(w)$, $c(w = 1)$, and D as independent degrees of freedom, which are minimized separately. The derivatives with respect to D need only be taken wherever D explicitly occurs; although the composition depends parametrically on D, the first variation of g with respect to the composition vanishes since $c(w)$ is taken as the minimum of g. This is true in the following section as well.

Pressure on the Walls

The net longitudinal pressure, Π_{zz} acting on the walls is the negative of the derivative of the total grand potential, G, with respect to the volume, V, as it is changed by increasing or decreasing the longitudinal dimension, D. We write this in terms of the grand potential per unit area, g and D as

$$\Pi_{zz} = -\frac{\partial G}{\partial V} = -\frac{1}{2}\frac{\partial g}{\partial D}$$

(5.107)

This yields

$$\Pi_{zz} = \frac{1}{2}\int_{-1}^{1} dw\left[\Pi_b[c(w)] - \Pi_b[c_b] + \frac{B}{2D^2}\left(\frac{\partial c(w)}{\partial w}\right)^2\right]$$

(5.108)

But the Euler-Lagrange equation satisfied by $c(w)$ implies

$$\frac{B}{2D^2}\left(\frac{\partial c(w)}{\partial w}\right)^2 = -\Pi_b[c(w)] + \Pi_b[c(0)] \qquad (5.109)$$

where $c(0)$ is the value of the concentration at the midplane. We thus find for the pressure:

$$\Pi_{zz} = (\Pi_b[c(0)] - \Pi_b[c_b]) \qquad (5.110)$$

where the osmotic pressure of the bulk is equal to $\Pi_b[c_b]$ and the osmotic pressure of the solute between the walls is equal to $\Pi_b[c(0)]$. As before, the pressure acting on the walls is related to the difference between the osmotic pressure of the system in the bulk and at the midplane. When the osmotic pressure of the system in the bulk is greater than that at the midplane, the net force on the walls tends to make the walls come closer and hence is attractive.

One can see that the force between the walls is attractive because in equilibrium, $\Pi_b[c_b]$ must be a maximum if the single-phase system is to be stable — *i.e.*, if there were another value of the composition with the same chemical potential, where $\Pi_b = \mu c_b - f_b[c_b]$ were the same or higher than $\Pi_b[c_b] = \mu c(z) - f_b[c(z)]$, the system would phase separate. (Note that minimizing the grand potential $f - \mu c$ implies that Π is maximized.) Thus, for a fixed chemical potential, $\Pi_b[c(0)] < \Pi_b[c_b]$ and $\Pi_{zz} < 0$, signifying an attractive interaction between the walls. As a specific example, consider the case where $f_b = \frac{1}{2}\chi c^2$. Equation (5.106) indicates that $\mu = \chi c_b$ so that

$$\Pi_{zz} = -\frac{1}{2}\chi(c(0) - c_b)^2 < 0 \qquad (5.111)$$

Polymeric Solutions

Equilibrium polymeric solutions (*e.g.*, long-chain polymeric molecules in a good solvent) also induce[9] attractive interactions between surfaces in a manner very similar to that described previously for small molecule solutions. Both the case of reversible and irreversible adsorption have been described in Ref. 9. The free energy of the bulk, similar to that of Eq. (5.103), is a functional not of the concentration, but of a variable, $\psi(\vec{r})$, whose square is proportional to the concentration of polymer. This can be understood in a heuristic manner by noting that the probability to find a given segment of a chain in space depends on the joint probability of *two* adjacent monomers along the chain (a more rigorous description can be found in Ref. 10). The surface interaction term is also similar to the one described above with the replacement of the surface concentration by the *square* of the variable, ψ. For reversible adsorption, where the polymer is

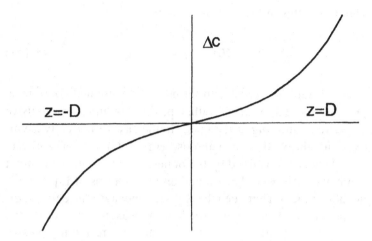

Figure 5.6 Concentration profile for perfectly asymmetric boundary conditions (*i.e.*, the solute is attracted to the right-hand wall and repelled by the left-hand wall) where $\Delta c = c(z) - c_b$, where c_b is the bulk solute concentration. The walls are located at $z = \pm D$.

in equilibrium with a bulk solution, the interaction between the plates is always attractive. For irreversible adsorption, the interaction between the surfaces can be repulsive.

Repulsive Interactions Between Surfaces

In the previous section, we showed that the presence of a solute in between two symmetric walls results in an effective interaction between the walls. This result depended on the fact that solution between the walls is in equilibrium with the bulk. However, there are situations when the interactions between the walls can be repulsive. This can occur when the boundary conditions at the two walls are very asymmetric (see Fig. 5.6) and therefore impose large gradients in the concentration at $z = 0$; the walls repel to reduce the effects of these large gradients.

This can be seen by considering Eq. (5.108) for the pressure. This expression is correct independent of the boundary conditions. However, the Euler-Lagrange equation, Eq. (5.109), is only correct for the case of symmetric boundary conditions where the derivative of the concentration profile is zero at $z = 0$. More generally, one can write the first integral of the Euler-Lagrange equation as

$$\frac{B}{2D^2} c_w'^2 = -\Pi_b[c(w)] + \Pi_b[c(0)] + \frac{B}{2D^2} c_w'^2(0) \qquad (5.112)$$

where $c_w = \partial c / \partial w$. Using this in Eq. (5.108), we find

$$\Pi_{zz} = \frac{B}{2} c_z^2(0) + (\Pi_b[c(0)] - \Pi_b[c_b]) \tag{5.113}$$

where we have now written the gradient in terms of the unscaled coordinate, $z = wD$. The pressure can therefore be either positive (signifying repulsive interactions) or negative (signifying attractions). In the case of perfectly asymmetric interactions with the wall, which impose $c(z = D) = c_b(1 + \delta)$ and $c(z = -D) = c_b(1 - \delta)$, where δ is fixed by the boundary conditions, one might expect that by symmetry, $c(0) = c_b$. The second and third terms of Eq. (5.113) cancel exactly and the interaction between the walls is repulsive (since the gradient squared term is positive). Thus, in the case of strongly asymmetric walls, the strong gradients in the concentration that must exist at $z = 0$ result in repulsive interactions that tend to reduce the effects of these gradients.

5.7 PROBLEMS

1. Van der Waals Interactions

Compute the van der Waals interaction energy for two membranes of thickness $2d$ whose centers are separated by a distance $2D$. What are the limiting values of the interaction for $D \gg d$ and for the case where the separation of the membranes is small compared to their thickness?

Compute the van der Waals interaction in Eq. (5.15) by transforming to relative coordinates: $Z = (z + z')/2$ and $\zeta = (z - z')/2$. The region of integration must be transformed carefully.

2. Dispersion Relations

Using Maxwell's equations and the appropriate boundary conditions as described in the text, derive the dispersion relations, Eq. (5.37) for the electromagnetic normal modes, for waves propagating in a one-dimensional system consisting of a dielectric film of thickness L and dielectric function $\epsilon_m(\omega)$ surrounded by two half-spaces of dielectric functions $\epsilon_1(\omega)$ and $\epsilon_2(\omega)$ respectively.

3. Osmotic Pressure and Electrostatics

What is the interaction energy as a function of separation of a periodic array of flat plates with a fixed charge density σ_0 and a spacing $2D$? Compute this explicitly from the free energy using the solution for $n(z)$. Find the longitudinal osmotic pressure and compare with osmotic pressure derived above: $\Pi_\ell = T n(0)$.

4. Interactions in the Debye-Hückel Approximation

What are the interactions between two plates with fixed charge density, σ, when the solution in the region between the plates includes the counterions and added salt that is in equilibrium with a bulk solution with concentration n_s? Use the Debye-Hückel approximation discussed in the text.

5. Electrostatics: A Free Energy Approach

In the text, we motivated the discussion of electrostatic interactions by considering the free energy as a function of the local charge density. This allows situations different from the simple ideal-gas limit to be considered since one can generalize the free energy by adding appropriate terms or degrees of freedom and minimize with respect to the charge density to find the equations that determine the local charge density. Consider the case where there is a specific interaction between the counterion molecules of the form

$$\int d\vec{r}d\vec{r}' \, U(\vec{r} - \vec{r}')n(\vec{r})n(\vec{r}') \tag{5.114}$$

where U is short-range. How would this affect the Poisson-Boltzmann equations?

Next, consider the case of counterions and added salt, where both the positive and negative salt ions are chemically distinguishable from the counterions. How does this change the free energy and the resulting Poisson-Boltzmann equations?

6. Surface Forces and Electrostatics

How does the result of Eq. (5.96) generalize if one includes interactions (e.g., excluded volume between finite-sized counterions) in the free energy? To see this, add a term to Eq. (5.85) that is $\frac{1}{2}Bn^2$ and attempt to repeat the calculation of the pressure.

7. Solute-Induced Interactions

Consider the specific case where the bulk free energy is given by a quadratic form, $f_b[c] = \frac{1}{2}\chi c^2$ and where the interaction with the wall has the form $f_w = -a_1 c(D)$. Find the concentration profile that minimizes the grand potential (i.e., the free energy with the constraint of conservation of solute) assuming that $c(D)$ is fixed. Next, using this solution for $c(z)$ compute the total (bulk and wall) Helmholtz free energy and minimize it with respect to $c(D)$ to find the concentration at the walls. (The minimization can either be done directly or using a method similar to that discussed in Chapter 4, Eq. (4.49).) Now, using this value of $c(D)$ evaluate the total Helmholtz free energy as a function of the wall spacing (consider the limiting cases of large and small separations) and discuss the nature and range of the interactions between the walls. Show that this interaction is independent of the sign of a_1 — it does not matter whether the walls are attractive or repulsive. Compare with the expression for the pressure acting on the walls derived in the text.

8. Pressure of Ideal Gas with Attractive Walls

Using Eq. (5.110) find the pressure on the walls in terms of $c(0)$ and c_b if the bulk free energy is that of an ideal gas. Show that the force between the walls is attractive.

9. Polymer Chain Between Attractive Walls

Consider a polymer chain between two attractive walls, separated by a distance $2D$. The free energy per unit area of the bulk solution is described[10] by a functional:

$$f_b = \frac{1}{2}B \int_{-D}^{D} dz \left(\frac{\partial \psi}{\partial z}\right)^2 \tag{5.115}$$

where the concentration of polymer, $c(z)$ (number of monomers per unit volume) is related[10] to ψ by $c(z) = \psi^2(z)$. Assume that the free energy describing the interactions with the walls is of the form:

$$f_w = -a_1 c(z) = -a_1 \psi^2 \tag{5.116}$$

The parameter $a_1 > 0$ for attractive interactions and $a_1 < 0$ for repulsive interactions of the monomers with the wall. By minimizing the free energy with respect to ψ subject to the constraint of conservation of the total number of monomers between the walls, find the total free energy as a function of the spacing D. Discuss the attractive or repulsive nature of the interaction between the walls as the parameter a_1 is varied. Find the pressure acting on the walls and compare to the general expression derived in the text. How does this case differ from that of solute-induced interactions for small molecules? For a discussion of interactions between plates for the case of a polymer in a poor solvent, see Ref. 11.

10. Attractive Interactions Between Walls

Consider a system where a medium is located between two walls separated by a distance $2D$. The normal to the walls is in the $\hat{z}$ direction and the plane $z = 0$ is the midplane between the walls. This medium is described by a *nonconserved* order parameter ψ that can be positive or negative and by a simple free energy per unit volume, f:

$$f = \int d\vec{r} \left(g[\psi] + \tfrac{1}{2} B |\nabla \psi|^2 \right)$$

where $g[\psi] = \tfrac{1}{2}\alpha\psi^2$. If the interactions with the wall are very strong, they determine a surface boundary condition. Consider the case where these boundary conditions result in $\psi = \psi_0$ at the right wall and $\psi = -\psi_0$ at the left wall.

Calculate the order parameter profile in the region between the walls. What is the effective interaction between the walls as a function of D?

What physical systems may present such behavior?

Consider the same problem but with the general local free energy, $g[\psi]$, where $g[0] = 0$. Write the first integral of the Euler-Lagrange equation for this general form and determine the integration constant in terms of the gradient of ψ at the midplane. Next, consider the free energy in scaled units where $w = z/D$

$$f = D \int_{-1}^{1} dw \left(g[\psi] + \frac{B}{2D^2} |\nabla \psi|^2 \right)$$

The change in f as D is varied can then be calculated explicitly:

$$\frac{\partial f}{\partial D} = \int_{-1}^{1} dw \left(g[\psi] - \frac{B}{2D^2} |\nabla \psi|^2 \right)$$

Use this expression and the first integral of the Euler-Lagrange equation to determine whether the interaction is attractive or repulsive in this more general case.

5.8 REFERENCES

1. J. N. Israelachvili, *Intermolecular and Surface Forces*, 2nd ed. (Academic Press, New York, 1992).

2. L. I. Schiff, *Quantum Mechanics* (McGraw-Hill, New York, 1968), ch. 8.

3. B. W. Ninham, V. A. Parsegian, and G. H. Weiss, *J. Statistical Physics* **2**, 323 (1970).

4. Yu. S. Barash and V. L. Ginzburg, in *The Dielectric Function of Condensed Systems*, eds. L. V. Keldysh, D. A. Kirzhnitz, and A. Maradudin, Vol. 24 in the series, *Modern Problems in Condensed Matter Sciences*, eds. V. M. Agranovich and A. A. Maradudin (North Holland, New York, 1989), p. 389.

5. L. D. Landau, and E. M. Lifshitz, *Statistical Physics*, 3rd ed., revised and enlarged by E. M. Lifshitz and L. P. Pitaevskii (Pergamon, New York, 1980), p. 159.

6. V. A. Parsegian, in *Annual Review of Biophysics and Bioengineering*, eds. L. J. Mullins, W. A. Hagins, and L. Stryer (Annual Reviews Inc., Palo Alto, CA, 1973) p. 221.

7. M. Elbaum and M. Schick, *Phys. Rev. Lett.* **66**, 1713 (1991); *J. Phys. I (France)* **1**, 1665 (1991).

8. L. D. Landau, and E. M. Lifshitz, *Theory of Elasticity*, 2nd ed., revised and enlarged (Pergamon, New York, 1970).

9. P. G. de Gennes, *Macromolecules* **15**, 492 (1982).

10. P. G. de Gennes, *Scaling Concepts in Polymer Physics* (Cornell University Press, Ithaca, NY, 1979).

11. K. Ingersent, J. Klein, P. Pincus, *Macromolecules* **19**, 1374 (1986).

CHAPTER 6

Flexible Interfaces

Fluid Membranes

6.1 INTRODUCTION

The previous chapter focused on the interactions between rigid surfaces or interfaces. However, some of the most interesting types of interfaces that occur in nature are not simple planar interfaces, but have an intrinsic roughness, often due to thermal effects, which can be important for two-dimensional systems such as membranes. In this chapter we consider the properties of flexible interfaces or membranes. We begin with a discussion of fluid membranes and the description of their free energy as a function of their curvature. Both phenomenological and more microscopic models are presented. We then describe the effects of thermal fluctuations on these nearly flat membranes. Finally, we consider the effects of interactions between fluctuating membranes and show how the fact that two membranes cannot interpenetrate leads to an effective, long-range repulsive interaction between them. In this chapter, we focus on the case of planar geometries; we assume that the molecules have self-assembled into a nearly flat membrane. The effects of curvature elasticity and thermal fluctuations on spherical and cylindrical systems are treated in Chapter 8, where the competition between different shapes in self-assembling molecules is discussed. The properties of the nearly flat membrane are important in understanding the bulk properties of self-assembling complex fluids such as microemulsions and vesicles as discussed, since these systems can be thought of as an ensemble of interacting membranes.

6.2   FLUID MEMBRANES AND SURFACTANTS

Membranes

In this chapter, we use the term membrane to denote a thin film of one material that separates two similar (bilayer membrane) or dissimilar (monolayer membrane) materials. We focus on fluid membranes (where there is no in-plane shear modulus and the only in-plane deformations are compressions/expansions), which are important in industrial applications such as encapsulation and cleaning. Furthermore, some fluid membranes are prototypes of biological systems, although it should be noted that true biological membranes often have several components and sometimes, even a solidlike underpinning that can give the membrane a shear rigidity.

Flexible, solid membranes, are also of interest. However, they are experimentally much less prevalent and are somewhat more complicated to treat since in addition to the membrane shape one must include the effects of shear. Their curvature energy is discussed in the problems at the end of this chapter. Another type of system that has received much theoretical attention is that of a tethered membrane[1], which may describe polymerized, but not crystalline sheets. While a single fluid membrane that is unconstrained by walls or other membranes is strongly affected by thermal fluctuations ("crumpled"), solid membranes, particularly if self-avoidance of the membrane is included, tend to be more weakly affected by fluctuations and are hence "flatter".

Amphiphiles

Surfactant molecules, or **amphiphiles** consist of molecules that combine both polar and nonpolar parts (see Fig. 6.1). Because of the hydrophobic interactions discussed in Chapter 5, these molecules tend to form *monolayer* films at polar-nonpolar (*e.g.,* water-oil) interfaces with the polar part of the molecule solvated in the water and the hydrocarbon part of the molecule in the oil. In this case, the properties of the film are, in general, not symmetric with respect to the interface. In a single solvent (*e.g.,* water), these molecules tend to form *bilayers* where the hydrocarbon parts of each monolayer are aggregated in the middle of the bilayer to reduce the contact between the water and the nonpolar parts of the molecule. When composed of a single species, the properties of such bilayer films are symmetric with respect to their two sides. Lipid molecules

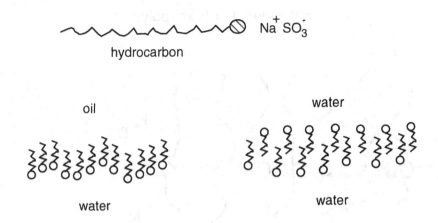

hydrocarbon

oil

water

water

water

Figure 6.1 Surfactant molecules self-assemble into monolayers separating water and oil and into bilayers separating two water regions.

are surfactant-like entities that generally have a polar head group and a double-chained hydrocarbon tail. They are important in biological applications. Another system that has amphiphilic properties are block copolymers, which consist of two immiscible polymers joined together by a covalent bond (see Fig. 6.2). If the two polymers are water soluble and oil soluble, respectively, these blocks are directly analogous to surfactants and thus form monolayer films at water-oil interfaces. Another type of interfacial activity exhibited by block copolymers is the tendency to form films at the interfaces between the two (immiscible) homopolymers that comprise the block. Such molecules are useful as compatibilizers of the two homopolymers and can be used to produce stable dispersions of one polymer in the other. This allows the formation of composite materials with particular properties which can be optimized by properly formulating the dispersion of the two types of polymers. (In the absence of the compatibilizer, the homopolymers would phase separate in equilibrium — *i.e.*, the dispersion would be unstable.)

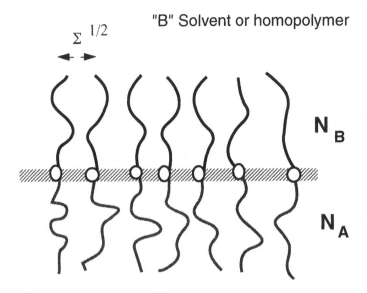

"B" Solvent or homopolymer

$\Sigma^{1/2}$

N_B

N_A

"A" Solvent or homopolymer

Figure 6.2 Block copolymers at the interface between a good solvent for the "A" blocks and a good solvent for the "B" blocks. The degree of polymerization of the two blocks are N_A and N_B respectively, and Σ is the area per molecule at the interface.

6.3   CURVATURE ELASTICITY OF FLUID MEMBRANES

Deformation Modes

Although fluid membranes can be composed of many different types of chemical and molecular species, their behavior (shapes, fluctuations, thermodynamics) can be understood from a unified point of view that considers the free energy of deformation of the membrane. If the membrane were constrained to lie in a plane, the only relevant energy would be the compression of the molecules — *i.e.,* change of the average area per molecule. This is analogous to sound waves in a three-dimensional fluid; there is no low-frequency response of the system to shear. However, since the membrane can also deform in the normal direction (out of the plane), there is an additional set of "modes" describing the conformations of the film. These out-of-plane deformations are known as bending or curvature modes and the free energy associated with such modes is known as the curvature free energy. For a membrane with finite thickness, we denote as pure curvature deformations those perturbations of the membrane that do not change the overall membrane volume, but where there may be local stretching and compression of different parts of the film. While a general deformation of the membrane involves both a change of volume and curvature, we shall see that the lowest energy deformations usually involve only the curvature. In most systems, changing the average volume of the membrane is a higher energy process and hence is less important when effects involving the thermal behavior of the membrane are considered. In addition, the location of the interface within a membrane of finite thickness can be chosen, to lowest order in the curvature energy, so that the surface which defines the interface undergoes no stretching or compression (neutral surface). However, except for a membrane that is perfectly symmetric about its midplane, the choice of this surface depends on the various elastic constants of the systems and is thus system dependent. For long-wavelength curvature deformations (whose wavelengths are much larger than the membrane thickness), the exact position of the interface within the membrane is not crucial.

Saturated Interfaces and Compressions

We consider a fluid, monolayer, membrane at a water-oil interface in equilibrium with a dilute solution of amphiphiles in the water and oil. In general, there is an equilibrium between those amphiphiles adsorbed at the interface and those in the bulk solution. For extremely small volume fraction of amphiphile, the surfactants will preferentially stay in solution due to their higher entropy of mixing with the solvent; the interface will have a relatively small number of amphiphiles adsorbed per unit area. However, this is not the case when the amphiphilic molecules are strongly insoluble in either solvent due to the unfavorable interactions of the polar groups with hydrocarbon solvents and of the hydrocarbon groups with polar solvents. The large energy cost of keeping these molecules in solution overcomes their entropy of mixing and at even moderately small volume fractions (which in practice can be very low $\sim 10^{-4}$ or less for surfactants, which strongly prefer the interface), the free energy cost for being in solution is too high and the amphiphiles will tend to accumulate at the interface.

As one increases the volume fraction of amphiphiles in the solution, more and more would go to the interface and the area per molecule, Σ, on the interface would decrease. However, the molecules cannot pack at infinite density at the interface. In the case where there exists a minimum in the packing energy of the flat interface at a value of $\Sigma = \Sigma_0$, the system will keep adding amphiphiles to the flat interface until Σ is reduced to a value close to Σ_0. If even more molecules are added to the system, instead of decreasing Σ further and thus *increasing* the free energy (since $\Sigma = \Sigma_0$ is a minimum), the amphiphiles will maintain their packing at $\Sigma \approx \Sigma_0$ and accommodate the extra molecules by creating *more* interface (*e.g.*, by rippling the flat interface or by incorporating oil into the water with the additional molecules located at the extra interface that is thus generated). When this happens, one says that the interface is saturated; instead of changing the packing area, the system accommodates more amphiphiles by making more interface under the condition of minimizing the free energy with respect to Σ. Of course, the interface may then have some curvature and the actual value of Σ may depend on the curvature (see below).

In general, one must consider the chemical potential of a molecule at the interface and in the solution. The equality of the two chemical potentials is the criterion for equilibrium and hence determines the area per molecule on the interface. When the amount of interface is fixed, as in the case of a single water-oil interface, this equality fixes Σ (see the problems at the end of Chapter 2). However, when the amount of interface can vary to minimize the free energy, Σ is determined[2] by minimizing the interfacial free energy per molecule; the chemical potential then determines the *number* of interfaces that exist in the system as well as the (small) volume fraction of surfactant that is not incorporated in these interfaces; the properties of *each* interface are determined to a first approximation by the minimization of the local free energy of the film.

The thermodynamics of these processes are discussed in detail in Ref. 2, where it is shown that there is a critical volume fraction of surfactant, ϕ^* above which there are many interfaces in the system, and the amount of surfactant not incorporated into these interfaces is small and remains approximately constant as the overall amount of surfactant, ϕ, is increased (ϕ^* is analogous to a critical micelle concentration — see Chapter 8). We therefore consider the simple case of surfactants that are strongly surface active (strongly insoluble in the bulk) so that at even very small volume fractions of amphiphile ($\phi^* \ll \phi \ll 1$), there are *many* interfaces (*e.g.*, vesicles, microemulsions) in the system in equilibrium. In this approximation, the fraction of surfactants in solution is very small and their volume fraction is approximately constant. The properties of the system are obtained by focusing on the properties of the interfaces. When, in addition, the interactions between these interfaces and their translational entropy can be neglected compared with the local deformation energies of the films, one can first minimize these local deformation energies to find the size and shape of the interfaces and then take into account the entropic and interaction effects as higher order corrections to the shape as determined by the curvature energy defined previously.

In addition to being characterized by the area per amphiphile, the interfacial membrane is also characterized by its thickness, λ, which can also change under deformations of the film. For simplicity, we assume that the equation of state of the flat membrane determines the thickness as a function of the area per molecule. (A simple example is the case of an incompressible molecule where the product of $\lambda\Sigma$ is constrained to equal the molecular volume so that $\lambda \sim 1/\Sigma$.) We thus take the flat membrane to be characterized only by the area per molecule, Σ; the curved membrane is characterized by both its curvature and area per molecule.

First consider a locally flat, isolated interface. Saturation occurs when the interfacial free energy achieves a minimum:

$$\frac{\partial f_0}{\partial \Sigma} = 0 \tag{6.1}$$

where f_0 is the free energy per molecule for a flat layer and Σ is the area per molecule. The free energy per molecule is minimized when $\Sigma = \Sigma_0$. The optimal value of the area per molecule arises from a balance of terms such as the entropy, and the interfacial-tension terms or attractions. The entropy favors a large area per molecule — because of the larger number of center of mass positions and chain conformations — while the interfacial-tension terms (*e.g.*, contact of the hydrocarbon chains with the water) and attractions favor a small value of Σ. Of course, there can be deviations in the area per molecule from this minimum and the energy cost of such a compression or expansion is

$$\Delta f_0 = \tfrac{1}{2} f_0''(\Sigma - \Sigma_0)^2 \tag{6.2}$$

where the primes signify a derivative with respect to Σ. However, these deformations are typically of higher energy than the curvature deformations; a membrane can change its shape or size with a much lower free energy cost than that required to compress or expand it. It is important to remember therefore, that for insoluble amphiphiles, it is the saturation of the interface and the minimization of the area per molecule that permits the usual surface-tension term to be neglected; the derivative $\partial f/\partial \Sigma = 0$. The surface tension is no longer relevant since the molecules adjust their area to optimize the free energy and it is therefore the curvature energy that mainly determines the properties of the film.

Curvature Deformations

We now consider a curved interface with principal curvatures κ_1 and κ_2 (see Chapter 1). The mean curvature is

$$H = \tfrac{1}{2}(\kappa_1 + \kappa_2) \tag{6.3}$$

and the Gaussian curvature is

$$K = \kappa_1 \kappa_2 \tag{6.4}$$

For now, we take the dividing surface from which the curvature is measured to be the polar-nonpolar interface. Below, we show that there exists a *neutral surface* where the stretching or compression modes are decoupled from the bending deformations; for the general surface treated here, we must consider this coupling. Although it complicates the problem somewhat, it is useful to treat the case of a general interface, since while the neutral surface is a convenient mathematical construction, microscopic models of amphiphilic interfaces often tie the interface location to a particular molecular site (*e.g.,* the bond joining the two parts of a block copolymer, the polar head group in a surfactant). The free energy per molecule is now a function of both Σ and the curvature. To describe the novel, large-scale structures observed in these systems and to characterize the low-energy deformations that are most strongly influenced by thermal fluctuations we consider radii of curvature whose length scales are much larger than molecular sizes. Specific molecular models for the free energy of the curved interface are considered in the following section. Here we show that one can obtain the form of the free energy as a function of curvature from very general considerations.

We write an expansion of the free energy per molecule, f, for small curvatures (the actual small parameter is the product of the curvature and the membrane thickness) up to second order in κ_1, κ_2. As explained in Chapter 1, the two invariants of the surface to this order in curvature are the mean and Gaussian curvatures; since the free energy of a fluid membrane must be invariant under rotations of the coordinate system, f is a function of H, H^2 and K to the order we consider. Thus,

$$f(\Sigma, H, K) = f_0(\Sigma) + f_1(\Sigma)H + f_2(\Sigma)H^2 + \bar{f}_2(\Sigma)K \tag{6.5}$$

where the coefficients of the curvature are, in general, functions of the equilibrium area per molecule, which itself may depend on curvature. The free energy of the flat film is f_0 and $f_1, f_2, \bar{f}_2$ are derivatives of the free energy with respect to H, H^2, K, respectively. Since the free energy of the flat layer has a minimum when $\Sigma = \Sigma_0$, a change in Σ that is *linear* in curvature, contributes a term *quadratic* in curvature to the free energy; the free energy has no term that is linear in $\Sigma - \Sigma_0$ because $\partial f_0 / \partial \Sigma_0 = 0$. We thus expand f_0 to second order and f_1 to first order in $\Sigma - \Sigma_0$ to find

$$f(\Sigma, H, K) \approx f_0(\Sigma_0) + \tfrac{1}{2}f_0''(\Sigma_0)(\Sigma - \Sigma_0)^2 + f_1(\Sigma_0)H$$

$$+f_1'(\Sigma_0)(\Sigma - \Sigma_0)H + f_2(\Sigma_0)H^2 + \bar{f}_2(\Sigma_0)K \tag{6.6}$$

where

$$f_0''(\Sigma_0) = \left(\frac{\partial^2 f_0}{\partial \Sigma^2}\right)_{\Sigma_0} \tag{6.7}$$

and

$$f_1'(\Sigma_0) = \left(\frac{\partial f_1}{\partial \Sigma}\right)_{\Sigma_0} \tag{6.8}$$

The terms proportional to H^2 and K are already quadratic order in our expansion. Their coefficients need only be kept to lowest order in the expansion of Σ and are therefore given by f_2 and $\bar{f}_2$ evaluated at $\Sigma = \Sigma_0$. Minimizing to find the equilibrium area per molecule of the *curved* interface, Σ^*, we find

$$\Sigma^* = \Sigma_0 - \left(\frac{f_1'}{f_0''}\right) H \tag{6.9}$$

Evaluating the free energy at the optimal value of the area per molecule, Σ^*, we see that f depends only on H and K. This defines the **curvature free energy** via

$$f(\Sigma^*, H, K) = g_0 + g_1 H + g_2 H^2 + \bar{g}_2 K \tag{6.10a}$$

where $g_0 = f_0(\Sigma_0)$, $g_1 = f_1(\Sigma_0)$, $\bar{g}_2 = \bar{f}_2(\Sigma_0)$, and

$$g_2 = f_2(\Sigma_0) - \frac{1}{2}\frac{f_1'^2}{f_0''} \tag{6.10b}$$

Notice that the correction term due to the fact that the area per molecule depends on curvature, is always negative since $f_0'' > 0$ by the minimization condition and the stability of the flat layer. Physically, this means that if the chains are allowed to adjust their area per molecule depending on the curvature, the monolayer will be less rigid upon bending than with a fixed area per molecule. We thus see that there is a term independent of curvature (the flat layer free energy at saturation), a term linear in curvature — which must vanish for a symmetric bilayer, but which is present for a monolayer — and a term quadratic in curvature.

Neutral Surface

In the preceding discussion, the area per molecule had a correction term due to curvature. This arose because of a coupling between the stretching and curvature of the surface — the term proportional to $(\Sigma - \Sigma_0)H$ in Eq. (6.6). This coupling can be eliminated by a shift in the normal direction of the surface of curvature by an amount λ whose magnitude is determined as follows. The curvature on the new interface changes (see the discussion of parallel surfaces in Chapter 1) according to

$$H' \approx H(1 + 2\lambda H) - \lambda K \qquad (6.11)$$

The higher order terms as well as the change in the Gaussian curvature are negligible if one keeps the free energy to second order in the curvatures only. The area per molecule defined with respect to the new interface, Σ' is related to the area per molecule defined on the original interface by

$$\Sigma' \approx \Sigma(1 - 2\lambda H) \qquad (6.12)$$

where we keep terms linear in H only since the energy Eq. (6.6) depends quadratically on deviations of Σ from its value on the flat interface. Rewriting the bending energy as a function of both Σ' and the curvatures (which have negligible, higher order corrections due to the shift of the interface position), and keeping terms up to order H^2, $(\Sigma - \Sigma_0)^2$, and $(\Sigma - \Sigma_0)H$, Eq. (6.6) becomes

$$f(\Sigma', H', K') \approx f_0(\Sigma_0) + \tfrac{1}{2}f_0''\,(\Sigma' - \Sigma_0)^2 + f_1(\Sigma_0)H'$$

$$+ H'(\Sigma' - \Sigma_0)\left[f_1' + 2\Sigma_0\lambda f_0''\right] + \bar{f}_2(\Sigma_0)K' + f_1(\Sigma_0)\lambda K'$$

$$+ \left[2f_0''\Sigma_0^2\lambda^2 + 2f'_1\Sigma_0\lambda + f_2(\Sigma_0) - 2f_1(\Sigma_0)\lambda\right]H'^2 \qquad (6.13)$$

where the extra terms arise from the change in the interface position.

The neutral surface is obtained by choosing the shift in the interface position, λ, so that there are no terms in $f(\Sigma, H, K)$ where the area and curvatures are coupled. From Eq. (6.13), one sees that this is the case when

$$\lambda = -\frac{f_1'}{2\Sigma_0 f_0''} \qquad (6.14)$$

The only dependence on Σ' in the free energy is a term proportional to $(\Sigma' - \Sigma_0)^2$, so that at the neutral surface, the minimum free energy configuration is given by $\Sigma' = \Sigma_0$; i.e., there is no change in the area per molecule compared with the flat surface. When the free energy is evaluated at the minimal value of Σ', only pure bending terms contribute and we find a free energy of the form of Eq. (6.10), with g_1 and $\bar{g}_2$ as given previously, but where g_2, the coefficient of H'^2, has several additional contributions (see the problems at the end of this chapter). Other choices of λ can be made, for example, to eliminate the term proportional to K'.

Curvature Energy

One can also discuss the curvature energy using symmetry considerations and relate it to the models analyzed previously. The most general form of the curvature free energy, f_c, *per unit area*, up to quadratic order in the two curvatures, κ_1 and κ_2 can be written in terms of the mean and Gaussian curvatures defined in Eqs. (6.3,6.4). One can write:

$$f_c = 2k\,(H - c_0)^2 + \bar{k}\,K \tag{6.15a}$$

which is equivalent to

$$f_c = \tfrac{1}{2}k\,(\kappa_1 + \kappa_2 - 2c_0)^2 + \bar{k}\,\kappa_1\kappa_2 \tag{6.15b}$$

This form for the free energy per unit area was discussed by Helfrich[3] and states that the mean curvature which minimizes the free energy has a value c_0, termed the **spontaneous curvature** of the membrane. The energy cost of deviating from the spontaneous curvature is the **bending or curvature modulus**, k. The parameter $\bar{k}$, known as the **saddle-splay modulus**, measures the energy cost of saddlelike deformations.

The spontaneous curvature describes the tendency of the surfactant film to bend toward either the water ($c_0 < 0$ by convention) or the oil ($c_0 > 0$). It is taken — in the absence of long-range interactions — to arise from the competition between the packing areas of the polar head and hydrocarbon tail of the surfactant molecules. If the interactions between the polar heads (as mediated through the intervening water and electrolyte) favor a smaller packing area than that dictated by the tail-oil-tail interactions, the surfactant film will tend to curve so that the heads (and the water) are on the "inside" of the interface. The bending moduli, k and $\bar{k}$, arise from the elastic constants determined by the head-head and tail-tail interactions. It is expected that these moduli are sensitively dependent on the surfactant chain length but only weakly dependent on the head-head interaction strength.

The Helfrich parameters, c_0, k and $\bar{k}$ can be derived from Eq. (6.10). Comparing Eqs. (6.10,6.15) and noting that f is an energy per molecule, while f_c is an energy per unit area, we identify

$$k = \frac{g_2}{2\Sigma_0} \tag{6.16a}$$

$$\bar{k} = \frac{\bar{g}_2}{\Sigma_0} \tag{6.16b}$$

$$c_0 = -\frac{g_1}{2g_2} \tag{6.16c}$$

This allows the curvature moduli to be obtained from the parameters of a given microscopic model that incorporates both the change in the area per molecule and the curvature. We note that a stable film will always have $k > 0$. However, the sign of $\bar{k}$ can be either positive or negative; films that prefer isotropic shapes (where the Gaussian curvature $K > 0$) such as spheres or planes will have $\bar{k} < 0$, while films that prefer saddle shapes (where the Gaussian curvature $K < 0$) will have $\bar{k} > 0$. One can show that the requirement that the quadratic term be positive definite implies that films are only stable if $2k + \bar{k} > 0$; otherwise higher order curvature terms are needed to stabilize the system.

A Simple Microscopic Model

As a simple microscopic model that will allow some more physical insight into the meaning of the curvature elastic moduli, we consider a monolayer of chains that we model as springs (see Fig. 6.3) with a spring constant, k_s, and with an equilibrium spring length, ℓ_s. We denote the actual (stretched or compressed) length of the spring by ℓ. We assume that the chains form an incompressible "melt" with no penetration of solvent into the chains. Their free energy is proportional only to the stretch of the springs; such a picture is applicable to polymeric molecules that pack incompressibly, but are stretched near an interface, so that their free energy is only due to their stretching. The area per chain at the interface is assumed to be fixed at a value Σ_0; in reality, this value is determined by the interactions that act on the polar head groups and our approximation assumes that these interactions are much stronger than the chain stretching energies, so that the optimal head area, Σ_0, determined by the interactions in the polar layer, is not affected by the chains. The energy per chain is thus

$$f = \tfrac{1}{2}k_s(\ell - \ell_s)^2 \tag{6.17}$$

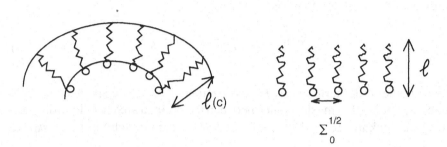

Figure 6.3 A curved surfactant monolayer of thickness $\ell(c)$.

and the incompressibility of the chains implies that the volume occupied by the layer is constant. For a flat layer, this would be written: $\Sigma_0 \ell = v_0$ where v_0 is the molecular volume. For a curved layer, the volume occupied by the chains depends on the curvature. The volume per molecule is

$$v_0 = \Sigma_0 \ell \left(1 + \ell H + \tfrac{1}{3}\ell^2 K\right) \qquad (6.18)$$

where H and K are the mean and Gaussian curvatures. (This formula is based on the integral of the expression for the area of a parallel surface, discussed in Chapter 1; it can also be derived by considering the cases of a sphere and a cylinder.)

Thus, the incompressibility condition relates the layer thickness, ℓ, and the area per molecule, Σ_0. For fixed Σ_0, this determines the free energy as a function of curvature obtained by solving Eq. (6.18) for ℓ and using this value for ℓ in Eq. (6.17). The result is

$$\ell = \ell_0 + \ell_1 H + \ell_2 H^2 + \ell_3 K \qquad (6.19)$$

where $\ell_0 = v_0/\Sigma_0$, $\ell_1 = -\ell_0^2$, $\ell_2 = 2\ell_0^3$, $\ell_3 = -\ell_0^3/3$. Note that for a flat layer, the incompressibility constraint determines the layer thickness to be equal to v_0/Σ_0; in general, this will not be equal to the thickness ℓ_s which minimizes the chain stretching energy. The flat layer is not, in general, the minimal energy state, implying that this monolayer has a spontaneous curvature related to the difference between the imposed thickness $\approx \ell_0$, and the preferred thickness, ℓ_s. Only when these two lengths are equal is the flat monolayer relieved of the frustration induced by the mismatch of these two lengths.

Using the incompressibility relation, we find that when $c_0 \ell_0 \ll 1$, the elastic energy per chain is given to lowest order by

$$f = \frac{k_s \ell_0^4}{2} \left[(H - c_0)^2 - \frac{2c_0 \ell_0}{3} K \right] \tag{6.20}$$

where a higher order term in $c_0 \ell_0 H^2$ has been neglected. The spontaneous curvature, c_0, is related to the difference between the optimal area per chain dictated by the head packing, Σ_0, and that preferred by the chain stretching energy, v_0 / ℓ_s:

$$c_0 = \frac{(v_0 - \ell_s \Sigma_0)}{\Sigma_0 \ell_0^2} \tag{6.21}$$

Equation (6.20) is equivalent to the "Helfrich" form of the curvature free energy by a simple transformation. The bending modulus (coefficient of H^2) and the saddle-splay modulus (coefficient of K) both increase as a power of the chain length. Of course, the spring constant, k_s, also depends on the equilibrium spring length, ℓ_s; a simple polymeric analogy yields in the limit of small curvatures, $k_s \sim 1/\ell_s \approx 1/\ell_0$. In that case, the bending modulus $k \sim \ell_s^3$. This variation of the bending modulus with the cube of the thickness is also characteristic of a bent, solid elastic plate as discussed in Ref. 4. There it is also shown that for an isotropic solid, the origin of the bending modulus is the shear modulus of the material.

The quantity c_0 is the spontaneous curvature of the membrane, which this model endows with a simple physical meaning: When the imposed head area, Σ_0 is larger than the optimal area, v_0 / ℓ_s, dictated by the chain packing, the preferred curvature is negative; the system prefers to pack with the heads on the "outside". Note that the free energy of the curved interface is lower than that of the flat interface; the system accommodates part of the strain induced by the mismatch between the heads and chains by bending.

Finally, we note that for a symmetric bilayer, composed of two monolayers, the curvature free energy is obtained by adding together the curvature energy of each monolayer (if interpenetration can be ignored). However, since each layer has a finite thickness, some care must be taken with the sign of the curvature of each monolayer and the location of the surface of curvature (see the problems at the end of this chapter). In the case that the spontaneous curvature of each monolayer is zero — *i.e.*, the amphiphile is balanced with respect to the packing of its heads and tails — the bending free energy per unit area of the bilayer has the simple form:

$$f_c = \tfrac{1}{2} k_b \, (\kappa_1 + \kappa_2)^2 + \bar{k}_b \, \kappa_1 \kappa_2 \tag{6.22}$$

where $k_b = 2k$ and $\bar{k}_b = 2\bar{k}$. This expression in correct in the approximation that the curvatures are small and corrections that take into account the fact that the two curvatures are not exactly equal and opposite can be ignored, since they result in higher order terms in f_c.

6.4 CURVATURE MODULI

For small curvatures, Eq. (6.15) shows that the curvature energy of a thin film is characterized by the three parameters k, $\bar{k}$, and c_0. The qualitative behavior of any system, including such properties such as the equilibrium shape, magnitude of thermal fluctuations, and any phase transitions, can of course be calculated as a function of these constants. However, the physics of the system can be radically different depending on the physical parameters; *e.g.*, a change in c_0 can induce shape changes in the system. It is thus of interest to relate the bending elastic moduli and the spontaneous curvature to the physics of the particular system of interest. This section first shows how these parameters are related to the pressure distribution in the membrane and then presents a simple but instructive microscopic model that relates k, $\bar{k}$, and c_0 to more molecular properties.

Relation to Pressure Distribution

The bending elastic moduli are determined by the curvature dependence of the free energy of the system *i.e.*, there is a resistance of the system to curvature. This curvature dependence is associated with a local area change; curvature changes the local area element. For an isotropic and homogeneous fluid, the work done in changing the volume is calculated using the relationship

$$\Delta F = - \int_{V_0}^{V} \Pi(V')dV' \tag{6.23}$$

where ΔF is the change in free energy due to an incremental, volume change from V_0 to V and $\Pi(V)$ is the local pressure (for a compressible system) or osmotic pressure (for a system of solvent and solute) against which this work is done. Usually Π is a function of V and Eq. (6.23) accounts for the total work that is done in expanding the system from V_0 to V; this requires a knowledge of Π at all volumes between V and V_0 and not just $\Pi(V_0)$. However, thin, liquid films that show a resistance to bending are anisotropic and one must consider separately the longitudinal pressure, Π_ℓ, which resists changes in the film thickness and the transverse pressure, Π_t, which resists changes in the film area. In the discussion of electrostatic interactions in Chapter 5, the longitudinal and transverse pressures were denoted by Π_{zz} and Π_{xx} respectively. In general, these quantities may vary within the film.

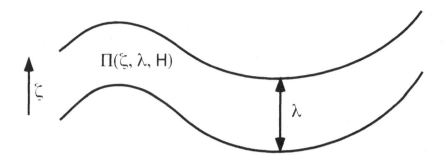

Figure 6.4 Membrane of thickness λ. The normal coordinate is ζ and the pressure varies in the $\hat{\zeta}$ direction and is also a function of curvature.

For solid films (*i.e.*, films with a shear modulus), there is an additional resistance to bending arising from the resistance to shear deformations[4]. This results in a nonzero curvature modulus even for a system which is elastically isotropic in the bulk. As shown below, this is not the case for systems with zero shear modulus; isotropic fluids show no resistance to bending deformations.

The curvature energy is essentially the work done in changing the *local* volume of the membrane related to the change in the area element due to the curvature; this work is done even if there is no *global* change in the volume of the membrane. We will therefore use a local version of Eq. (6.23) to calculate the curvature energy and consider the volume change due to curvature. This method for calculating the bending energy was introduced in Refs. 5,6 where the area change was considered.

We consider an infinitely large membrane of thickness λ and area per molecule Σ (defined on the "bottom" ($\zeta = 0$) plane of the membrane shown in Fig. 6.4) and calculate the curvature free energy as a function of these pa-rameters that characterize the global properties of the membrane. In practice, both Σ and λ are functions of curvature. However, it is simplest to first keep these parameters fixed and derive the curvature energy expansion; this requires that we know Π for any given Σ and λ. Afterward, one can minimize the free energy with respect to these parameters (equivalent to determining an equation of state) and/or use constraints of constant volume to constrain either or both of them. (See the previous section for an example of how the equilibration of the area per molecule can affect the bending modulus.) For example, the thickness of the membrane can be found by requiring that the curvature energy include only pure curvature deformations — *i.e.,* that there be no overall compression or expansion of the membrane. Here, we make this choice and determine the film thickness of the curved film, λ, by requiring that the total volume of the system is kept constant (enforcing global incompressibility).

For constant, or slowly varying curvatures, the area element changes due to the curvature. The equality of the volumes of the flat and curved membranes is written (see Chapter 1):

$$\Sigma \int_0^\lambda d\zeta \left(1 + 2\zeta H + \zeta^2 K\right) = \Sigma_f \, \lambda_f \tag{6.24}$$

where λ_f is the thickness and Σ_f is the area per molecule of the flat membrane. The coordinate ζ is the distance from the bottom plane of the membrane in the normal direction, and H and K are the mean and Gaussian curvatures respectively. To take into account the effects of the change in the area per molecule we define

$$\lambda_0 = \lambda_f \, \frac{\Sigma_f}{\Sigma} \tag{6.25}$$

We therefore include all the dependence of the parameters on the actual area per molecule of the curved membrane, Σ, in their behavior as a function of λ_0 and rewrite Eq. (6.24) as

$$\int_0^\lambda d\zeta \left(1 + 2\zeta H + \zeta^2 K\right) = \lambda_0 \tag{6.26}$$

As mentioned previously, once the curvature energy expansion is obtained in terms of λ_0, it can further be minimized over Σ and the "area-equilibrated" expansion can be derived as was done in Eqs. (6.5,6.9,6.10). This implies that under the constraint of constant local volume, the thickness of the curved membrane, λ, is related to that of the flat membrane, λ_0, by

$$\lambda \left(1 + \lambda H + \frac{K\lambda^2}{3}\right) = \lambda_0 \tag{6.27}$$

To second order in the curvatures, this implies that

$$\lambda \approx \lambda_0 \left(1 - \lambda_0 H - \frac{K\lambda_0^2}{3} + 2\lambda_0^2 H^2\right) \tag{6.28}$$

We assume that the curvatures are slowly varying in the plane of the membrane (or are constant as they are for cylindrical and spherical curvature) so that the pressure depends only on ζ. The surface of curvature is measured from the bottom of the membrane located at $\zeta = 0$. We calculate f_c, which is the curvature free energy per unit area of the membrane base, A_0, (which we hold fixed for now) from the change in free energy arising from the change in the local area element, $A_0(2\zeta H' + \zeta^2 K')$, due to the curvatures H' and K'. Using the principle of virtual work to compute the free energy, we note that the work is independent of the path chosen. Thus, we choose to first keep the thickness fixed at λ_0 and change the area elements commensurate with the curvature. Since we will need contributions up to second order in the curvature, we consider the work done in continuously changing the mean curvature from $H' = 0$ to $H' = H$ and the Gaussian curvature from $K' = 0$ to $K' = K$; the mean and Gaussian curvatures are two independent degrees of freedom of the membrane as discussed in Chapter 1. This requires an integration over the differential changes in curvature. We then have a curved membrane of thickness λ_0. For this process of local area change, it is the *transverse pressure*, Π_t, that does work against the differential volume change; the initial volume element, $A_0(1 + 2H'\zeta + K'\zeta^2)d\zeta$, is modified because of the change in the area element upon increasing H' by dH' and K' by dK':

$$dV = A_0 \left(2\zeta \, dH' \, d\zeta + \zeta^2 \, d\zeta \, dK' \right) \tag{6.29}$$

Next, we compute the work needed to change the thickness of this curved membrane thickness from λ_0 to λ. Here, the membrane surface at $\zeta = \lambda'$ does work against the change in the thickness and we compute this work by multiplying the *longitudinal* pressure, Π_ℓ, at the top surface by the volume element $d\lambda'$ and integrating λ' from λ_0 to λ. We therefore write:

$$f_c = -f_a - f_t \tag{6.30a}$$

where

$$f_a = \int_0^H dH' \int_0^{\lambda_0} d\zeta \, 2\zeta \, \Pi_t(\zeta, \lambda_0, H', K')$$

$$+ \int_0^K dK' \int_0^{\lambda_0} d\zeta \, \zeta^2 \, \Pi_t(\zeta, \lambda_0, H', K') \tag{6.30b}$$

$$f_t = \int_{\lambda_0}^{\lambda} d\lambda' \, \Pi_\ell(\lambda', \lambda', H, K)(1 + 2H\lambda' + K\lambda'^2) \tag{6.30c}$$

where $\Pi_t(\zeta, \lambda_0, H', K')$ is the local, ζ dependent, transverse pressure in the membrane of thickness λ, which is bent with the mean and Gaussian curvatures H', K' respectively. The longitudinal pressure, Π_ℓ, is also a function of the thickness and the curvature. The term labeled f_a is the work due to the change in the area element at fixed $\lambda = \lambda_0$ and the term labeled f_t is the work due to the change in thickness of the already bent membrane. This form for f_c guarantees that the curvature energy is calculated relative to that flat state — *i.e.*, if $H = K = 0$ (and thus $\lambda = \lambda_0$), the bending energy must vanish. More importantly, Eq. (6.30) has the property that any constant, isotropic terms in the pressures do *not* contribute to the curvature energy, f_c. This can be seen explicitly by using Eq. (6.27) for λ in Eq. (6.30) with $\Pi_t = \Pi_\ell$ equal to a constant. The physical reason for this is that the curvature energy requires an interaction that extends throughout the thickness of the membrane; an ideal gas or small molecule fluid with constant density and hence constant $\Pi_t = \Pi_\ell$ takes the shape of its container and in the absence of compression or expansion, guaranteed by Eq. (6.27), has zero curvature energy.

In principle, to evaluate the curvature energy we must know the pressures at all values of the intermediate curvatures H', K'. However, to find f_c to second order in the curvature, it is sufficient to expand each of the pressures, Π_i ($i = t, \ell$), to first order in H and we write:

$$\Pi_i(\zeta, \lambda, H', K') \approx \Pi_{i0}(\zeta, \lambda) + \left(\frac{\partial \Pi_i}{\partial H'}\right)_{H'=0} H' + \dots \qquad (6.31)$$

where $\Pi_{i0}(\zeta, \lambda)$ is the local transverse ($i = t$) or longitudinal ($i = \ell$) pressure of a membrane with thickness λ in the flat state. Eq. (6.31) requires that we solve for the free energy of the membrane in the curved state; *i.e.*, we must solve the entire problem to linear order in curvature, obtain the pressure and then take the curvature derivative. We use this expansion in both f_a and f_c and perform the integrals over the dummy curvature variables in f_a. The contribution from f_t is evaluated by expanding the integral for small values of $\lambda - \lambda_0$ (see Eq. (6.28)) to second order. Keeping terms up to second order in curvature, we find

$$f_c = -\int_0^{\lambda_0} d\zeta \left[\Pi_{t0}(\zeta, \lambda_0) (2\zeta H + \zeta^2 K) + \left(\frac{\partial \Pi_t}{\partial H}\right)_{H=0} H^2 \zeta \right]$$

$$-\lambda_0 \left(-\lambda_0 H - \frac{K\lambda_0^2}{3}\right) \Pi_{\ell 0}(\lambda_0, \lambda_0) + \left(\lambda_0^2 H^2 \left(\frac{\partial \Pi_\ell(\lambda_0, \lambda_0)}{\partial H}\right)_{H=0}\right)$$

$$-\frac{1}{2} \left(\lambda_0^4 H^2 \left(\frac{\partial \Pi_\ell(\lambda', \lambda')}{\partial \lambda'}\right)_{\lambda'=\lambda_0}\right) \qquad (6.32)$$

Comparing powers of H, H^2, and K in Eqs. (6.15,6.27,6.32) allows us to identify the curvature moduli as moments of the pressure distribution. Thus,

$$kc_0 = \frac{1}{2} \int_0^{\lambda_0} \tilde{\Pi}_0(\zeta, \lambda_0)\, \zeta\, d\zeta \tag{6.33a}$$

$$\bar{k} = -\int_0^{\lambda_0} \tilde{\Pi}_0(\zeta, \lambda_0)\, \zeta^2\, d\zeta \tag{6.33b}$$

$$k = -\frac{1}{2} \int_0^{\lambda_0} \left(\left[\frac{\partial \tilde{\Pi}_0}{\partial H}\right]_{H=0} - \left(\left[\frac{\partial \Pi_\ell(\lambda_0, \lambda_0)}{\partial H}\right]_{H=0} + \left[\frac{\partial \Pi_\ell(\lambda, \lambda)}{\partial (1/\lambda)}\right]_{\lambda=\lambda_0} \right) \right) \zeta\, d\zeta \tag{6.33c}$$

where

$$\tilde{\Pi}_0(\zeta, \lambda_0) = \Pi_{t0}(\zeta, \lambda_0) - \Pi_{\ell 0}(\lambda_0, \lambda_0) \tag{6.33d}$$

If the pressure field is continuous in space, then the thickness, λ_0 can be set to infinity and $\Pi_{\ell 0}(\lambda_0, \lambda_0) = 0$, corresponding to a zero pressure boundary condition far away from the membrane. If the pressure is nonzero in only a finite region (*e.g.*, a fluid or a gas contained between two walls with a finite thickness), the dependence on the difference in pressures in Eq. (6.33) guarantees that there is no curvature energy for an isotropic, fluid system with $\Pi_t = \Pi_\ell$ constant. (In this case, our expressions differ somewhat from those of Refs. 5,6.) For a solid, however, the shear response of the system results in a nonzero bending modulus, even for an isotropic elastic medium[4]. We note that the combination kc_0 and the saddle-splay modulus $\bar{k}$ are simply related to the moments of the pressure distribution of the flat membrane, while the bending modulus itself requires that the change in pressure due to curvature be calculated to linear order in H. For a given microscopic model (*e.g.*, charged membranes, polymers at an interface), this requires a solution of the density profile and the resulting free energy and pressure in the curved geometry. However, k and $\bar{k}$ often scale in an identical manner with the microscopic parameters (*e.g.*, charge density, membrane thickness); one can therefore find $\bar{k}$ quite simply and infer that k scales similarly. Finally, we note that for membranes with stress-free boundaries at $\zeta = \lambda$, (*i.e.*, $\Pi_\ell(\lambda, \lambda) = 0$ for all values of λ), and the expression for k simplifies.

Example: Electrostatic Pressure and Curvature Energy

Calculate the quantities kc_0 and $\bar{k}$ as functions of the interlayer distance due to the combined entropy and electrostatic energy of the charges in a periodic stack of charged surfaces, with a charge per unit area, σ_0. Each fixed charge contributes to the solution a counterion that is soluble in the intervening electrolyte and we consider the case of no added salt (see Chapter 5). The spacing of the surfaces in the stack is $2D$. (Although k is more difficult to compute, one expects it to scale like $\bar{k}$, but to have a positive sign.) Assume that all the surfaces bend in phase.

For a periodic stack, we need only consider the bending within a unit cell in order to consider modes where all the membranes bend in phase. The unit cell we consider consists of one charged plane at $\zeta = 0$, surrounded by its gas of counterions for a distance $\zeta = \pm D$. Thus, the center of curvature is referred to this plane. The expressions for the bending moduli in Eqs. (6.33) involve moments of the pressure distribution — i.e., integrals with powers of ζ. Thus, Π_i $(i = t, \ell)$ at $\zeta = 0$ does not contribute to the bending coefficients, and we need the pressure distribution for the counterions. (If we had taken the center of curvature elsewhere, we would also have to include the pressure due to the fixed charge layer.) In Chapter 5, we computed the transverse and longitudinal pressures for the flat film where the fixed charges are located at $z = \pm D$ and the midplane is at $z = 0$. To transform from this coordinate system to the one used to calculate the bending, we identify $z = \zeta - D$ with the region above the surface of curvature and $z = \zeta + D$ with the region below. Also in Chapter 5, we showed that the pressures are simple functions of the charge density which varies in space as

$$n = \frac{n_0}{\cos^2 k_0 z} \tag{6.34}$$

where the boundary conditions fix

$$k_0^2 = 2\pi n_0 \ell \tag{6.35}$$

and

$$n_0 = \frac{k_0 \sigma_0}{2 \tan k_0 D} \tag{6.36}$$

where σ_0 is the fixed charge per unit area on the membranes (co-ion charge) and $\ell = e^2/(\epsilon T)$ is the Bjerrum length. In the high charge limit, $k_0 D \approx \pi/2$ while in the low charge limit where the counterions are a nearly uniform gas, $k_0 D \ll 1$.

To compute the bending coefficients from Eq. (6.33), we recall that they are only functions of $\tilde{\Pi}_0$ defined in that equation. Therefore, those parts of $\tilde{\Pi}_0$ that are independent of ζ do not contribute to $\bar{k}, k, c_0$. Using the expressions for $\Pi_{t0} = \Pi_{xx}$ and for $\Pi_{\ell 0} = \Pi_{zz}$ from Chapter 5, we find

$$\tilde{\Pi}_0 = 2T\left[n(z) - n(0)\right] \tag{6.37}$$

where we note that the system is neutral within the unit cell. From Eq. (6.33) (integrated with respect to the coordinate ζ) we see that $c_0 = 0$; the symmetry of the system requires that there be no spontaneous curvature. The saddle-splay constant, $\bar{k}$ is given by Eq. (6.33) and using the symmetry we find

$$\bar{k} = -4Tn_0 \int_0^D d\zeta \, \zeta^2 \left[\sec^2\left(k_0(\zeta - D)\right) - 1\right] \tag{6.38}$$

For high charge densities and large separations, $k_0 D \approx \pi/2$ so that the integral can be approximately written as

$$\bar{k} = -4Tn_0D^3 \int_0^1 dx \, x^2 \left[\csc^2\left(\frac{\pi x}{2}\right) - 1\right] \tag{6.39}$$

From Eq. (6.36) for n_0 with $k_0 D \approx \pi/2$, we find $n_0 \approx \pi/(8D^2\ell)$ so that

$$\bar{k} \approx -T\frac{\beta\pi D}{2\ell} \tag{6.40}$$

where $\beta \approx 0.23$ is the dimensionless integral in Eq. (6.39). Thus the Gaussian bending modulus is negative, signifying a tendency to make isotropic curvatures. The magnitude of $\bar{k}$ increases linearly with the spacing between the layers. Similar scaling for the bending modulus has been discussed in Ref. 7. The relationship between the bending moduli and the pressure distribution greatly simplifies the calculation and may prove very useful in more complex systems.   ■

6.5   FLUCTUATIONS OF FLUID MEMBRANES

Height and Normal Fluctuations

We consider a single membrane with zero spontaneous curvature, described by the bending free energy of Eq. (6.22). In the Monge gauge, in the approximation of small curvatures, the free energy per unit area can be written (see Chapter 1):

$$f_c = \tfrac{1}{2} k \left(h_{xx} + h_{yy} \right)^2 \tag{6.41}$$

where $h(x, y)$ describes the height of the membrane and $h_x = \partial h / \partial x$. In Fourier space ($\vec{q} = (q_x, q_y)$), this becomes

$$f_c = \frac{1}{2A} \, k \sum_{\vec{q}} q^4 |h_q|^2 \tag{6.42}$$

where A is the area in the $x - y$ plane. Note that the free energy, which is proportional to q^4 per mode, is much "softer" at long wavelengths than that corresponding to surface-tension fluctuations (whose energies are proportional to q^2, as shown in Chapter 3). We thus expect the effect of thermal fluctuations on these membranes to be even more significant than for the problem of thermal roughening in the presence of surface tension.

Strictly speaking, this free energy should be augmented by a Lagrange multiplier that accounts for the fact the membrane is composed of a fixed number of amphiphiles which, to a good approximation, is equivalent to imposing a fixed total area. Thus, one might imagine that there should be an additional surface-tension-like term to account for this area conservation. However, the Lagrange multiplier or surface-tension-like term in this case, must vanish as the area becomes infinite and this constraint is unimportant in this limit (see the discussion of microemulsions in Chapter 8, where these constraints do become important). In addition, for a finite-size membrane the extra area introduced by the "crumpling" due to thermal fluctuations is small if the membrane is smaller than a persistence length (defined in the next subsection). For the solid or fluid surface discussed in Chapter 2 — with no amphiphile to stabilize the interface — the surface tension, related to the difference in free energy between a molecule in the bulk and a molecule at the interface, is a microscopic property of the material and is unrelated to the size of the surface. In our case, the surfactants prefer the interface over the bulk and we have a large number of interfaces. The amphiphiles actually choose their area per molecule to minimize their free energy as discussed previously. Thus, the effective surface tension can be taken to be zero, since for a flat interface, saturation implies that the change in free energy per molecule as a function of the area per molecule is zero:

$$\frac{\partial F}{\partial A} \sim \frac{\partial F}{\partial \Sigma} \sim 0 \tag{6.43}$$

It is only the weak, global conservation of area constraint that gives a term in the free energy proportional to the area. This constraint is negligible for large areas and we thus consider here the thermodynamic limit of the fluctuations of infinitely large, nearly flat membranes.

By the equipartition theorem,

$$\left\langle |h_q|^2 \right\rangle = \frac{T}{kq^4} \tag{6.44}$$

The mean-square fluctuation of the height increases algebraically with the system size;

$$\left\langle h(\vec{r})^2 \right\rangle = \frac{1}{A} \sum_{\vec{q}} \left\langle |h_q|^2 \right\rangle \sim \frac{T}{k} L^2 \tag{6.45}$$

for the surface tension dominated two-dimensional interface, it only diverged logarithmically. For the membrane, a quantity of interest is the normal-normal correlation function:

$$g_n(\vec{r}) = \left\langle (\hat{n}(\vec{r}) - \hat{n}(0))^2 \right\rangle \tag{6.46}$$

where for small curvatures, the normal is related to the height by

$$\hat{n} \approx \hat{z} - h_x \hat{x} - h_y \hat{y} \tag{6.47}$$

It is this correlation function that defines the curvature of the membrane since it describes how the normal bends as one goes along the membrane a distance $\vec{r}$. The correlation function is given by

$$g_n(\vec{r}) = \frac{2}{A} \sum_{\vec{q}} q^2 \left\langle |h_{\vec{q}}|^2 \right\rangle (1 - \cos \vec{q} \cdot \vec{r}) \tag{6.48}$$

where A is the membrane area. As $r \to \infty$ the integral diverges logarithmically; it is evaluated by using

$$\int_0^{2\pi} \cos(qr \cos \theta) d\theta = 2\pi J_0(qr) \tag{6.49}$$

for the angular integral (where J_0 is the zeroth order Bessel function) and approximating the remaining integral. The result is that for large distances:

$$g_n(\vec{r}) \approx \frac{\alpha T}{4\pi k} \log \frac{r}{a} \tag{6.50}$$

where a is a microscopic distance and α is a constant of order unity.

Persistence Length and Area Renormalization

The normal-normal correlation function can be used to define the persistence length of the membrane[8] as the distance over which the normal becomes decorrelated via the thermal undulations: the distance r at which $g_n(r)$ is of order unity. The persistence length, ξ_k, is defined as

$$\xi_k = a \exp \left[\frac{4\pi k}{\alpha T} \right] \tag{6.51}$$

Note that ξ_k is exponential in the curvature modulus. Another interpretation of the persistence length is that for length scales below ξ_k, the membrane is locally flat, while for larger length scales, it executes a random walk in space, much like the two-dimensional analogue of a polymer. At length scales of order the persistence length, the curvature energy competes with the entropy of the membrane conformations. This effect has also been[9,10] expressed in terms of the effective bending modulus of the membrane as a function of length scale. The authors of Refs. 9,10 show that the *effective* bending modulus, k_e depends on the length scale of the patch L, as

$$k_e(L) = k \left[1 - \frac{\alpha T}{4\pi k} \log(L/a) + ... \right] \tag{6.52a}$$

where α is a constant of order unity. When $L \approx \xi_k$ the effective modulus becomes small. These effects are related to the fact that the membrane area is larger than that of its projection due to the thermal ripples. A similar calculation gives the excess area, ΔA in a patch of membrane of projected area, $A = L^2$, due to the thermal rippling:

$$\frac{\Delta A}{A} \approx \frac{T}{4\pi k} \log(L/a) \tag{6.52b}$$

Persistence Length and External Potential

In some cases a membrane may be subject to an external potential, characterized by a strength V_0, which couples to the film position. This can occur in a many membrane system where the external potential is just the effective "mean-field" of the intermembrane interactions; another example is the force of gravity. In this case, the fluctuations described previously are no longer divergent (see the discussion of the rough interface under gravity in Chapter 3). The free energy is augmented by a term of the form: $V_0 \int d\vec{r} h^2(\vec{r})$ and the fluctuation spectrum obeys

$$\langle |h_q|^2 \rangle = \frac{T}{k} \frac{1}{\left(q^4 + \xi_0^{-4}\right)} \tag{6.53}$$

where the length at which the small q fluctuations are cut off is given by $\xi_0 \sim (k/V_0)^{1/4}$.

6.6   INTERACTIONS OF FLUID MEMBRANES

Swollen Membranes

A lamellar, lyotropic liquid crystal can be described as a stack of fluid membranes in a common solvent — *e.g.*, bilayer membranes in water. In some systems, the average spacing between membranes, d, can be large — thousands of Å, while in others, the average spacing has a maximum value of only several tens of Å; trying to swell the system further by the addition of solvent results in phase separation of a membrane phase and excess solvent. Obviously, the "unbound" systems that can be swollen to large distances have only negligible attractive interactions, while the tightly bound systems are affected by the van der Waals interactions, which tend to decrease like $1/d^4$ (see Chapter 5). While charged systems at low salt concentrations can have strong electrostatic repulsions that can stabilize phases with large values of d, all membrane stacks are subject to another repulsive interaction arising from entropic restriction which is long-range and sometimes termed the "Helfrich" interaction[11].

Steric Repulsion of Membranes

To calculate the effect of entropic restriction in a many membrane system, we consider a system with membranes whose average positions lie on a one-dimensional lattice along the $\hat{z}$ direction, with lattice spacing d as shown in Fig. 6.5. A given membrane has an absolute height $Z_n = nd + h_n(x, y)$, where n is an integer which indexes the lattice positions and $h_n(x, y)$ is the local (*i.e.*, it may vary in the xy plane), deviation of the membrane from its average value,

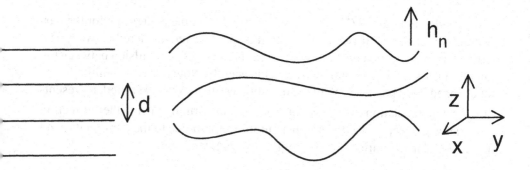

Figure 6.5 A stack of membranes at an average spacing d but with randomness due to fluctuations.

$\langle Z_n \rangle = nd$. If there are no fluctuations of the membranes, then $h_n(x, y) = 0$. Due to the fluctuations, the membranes collide and lose entropy in these collisions. This can be understood if one considers the case of hard-core repulsions between membranes; the excluded volume of the neighboring membranes limits the configuration of any given membrane, thus reducing its entropy. This entropic limitation implies that the free energy per membrane of a stack of membranes must be greater than that of a single, free film.

Following Helfrich[11] we assume that the net result of these collisions is that each membrane experiences an effective interaction with its nearest neighbors which has as its lowest *energy* state, the periodic configuration, where

$$Z_{n+1} - Z_n = d \tag{6.54}$$

This interaction is represented as a quadratic form in the deviation of $(Z_{n+1} - Z_n) - d$ from zero, or equivalently in $h_{n+1}(x, y) - h_n(x, y)$. In addition, there is the bending energy of each membrane. The Hamiltonian, $\mathcal{H}$, is thus written:

$$\mathcal{H} = \int dx\, dy\, u(x, y) \tag{6.55a}$$

$$u = \tfrac{1}{2} B \sum_n (h_n - h_{n+1})^2 + \tfrac{1}{2} k \sum_n \left(h_{n_{xx}} + h_{n_{yy}} \right)^2 \tag{6.55b}$$

Here k is the bending modulus, and the subscripts of xx and yy represent two derivatives of the membrane position variable — *i.e.*, $\tfrac{1}{2} \left(h_{n_{xx}} + h_{n_{yy}} \right)$ is the mean curvature of the nth membrane. This expression is correct for membranes with gentle undulations ($\nabla h \ll 1$); otherwise the simple expression for the curvature is incorrect and the area constraints must be reconsidered as well. The compressional elastic constant, B, represents an effective repulsion between the membranes and will be computed self-consistently. Note that this Hamiltonian is unchanged if the positions of all the membranes are uniformly shifted, representing a trivial translation of the system. Fourier transforming in both the $\hat{z}$ direction (Fourier wavevector Q with an upper cutoff of $\pm \pi/d$ due to the periodicity) and the $x - y$ plane (Fourier wavevector $\vec{q} = (q_x, q_y)$) we have

$$\mathcal{H} = \sum_{\vec{q},Q} |h(\vec{q}, Q)|^2 \left[B(1 - \cos Qd) + \tfrac{1}{2}kq^4 \right] \tag{6.56}$$

Free Energy of Undulations and the Repulsive Interaction

The Boltzmann factor corresponding to the Hamiltonian of Eq. (6.56) is a Gaussian, so the free energy, F is easily evaluated from

$$F = -T \log \left[\prod_{\vec{q},Q} \int dh(\vec{q}, Q) \, e^{-\mathcal{H}/T} \right] \tag{6.57}$$

Performing the integral, we find that the difference in free energy per unit volume, Δf, between the many membrane system and that of a single membrane (where $d \to \infty$ and we anticipate that $B \to 0$) is given by

$$\Delta f = \frac{T}{2(2\pi)^3} \int d\vec{q} \, dQ \, \log \left[\frac{B(Q) + kq^4}{kq^4} \right] \tag{6.58}$$

where $B(Q) = 2B(1 - \cos Qd)$. Integrating first over $\vec{q}$ with the upper cutoff set to ∞, we find

$$\Delta f = \frac{T}{16\pi} \int_{-\pi/d}^{\pi/d} dQ \, \sqrt{\frac{2B(1 - \cos Qd)}{k}} \tag{6.59}$$

This integral can be written:

$$\Delta f = \frac{T}{2\pi d} \sqrt{\frac{B}{k}} \tag{6.60}$$

Now, the modulus B is related to the second derivative of the free energy with respect to the average layer spacing; *i.e.*, imagine a uniform expansion or compression of the system. The restoring force is just the effective value of B which is proportional to the macroscopic compressibility of the system. Thus, following Helfrich[11], we can obtain a self-consistent equation to determine B from

$$B = \frac{\partial^2(\Delta f d)}{\partial d^2} \tag{6.61}$$

Using Eq. (6.60) in Eq. (6.61), we find that

$$B = \frac{9T^2}{\pi^2 k} \frac{1}{d^4} \tag{6.62}$$

and the free energy difference per unit (projected) area, $\Delta f_a = \Delta f\, d$, is

$$\Delta f_a = \frac{3T^2}{2\pi^2 k} \frac{1}{d^2} \tag{6.63}$$

representing an effective repulsion that decays slowly as d is increased. If one approximates Eq. (6.59) by its small Q expansion, one obtains an expression for the free energy per unit area which has the same scaling as Eq. (6.63) but with a numerical coefficient of $3\pi^2/128$ as in Ref. 11. This long-range, entropic repulsion is present in all multimembrane systems. In addition to this repulsion, the specific Coulomb repulsions due to charge effects and/or attractions due to van der Waals interactions may result in an effective attractive well that "binds" the membranes at a particular distance. In the absence of such attractions, the membranes can be swollen to large distances. However, at very large swellings, the one-dimensional stacking order may "melt" and the membranes may form a disordered, bicontinuous, spongelike phase.

6.7 PROBLEMS

1. Bending of Solid and Liquid Films

Solid thin films also offer resistance to bending and their energy can therefore be written in the form of a curvature expansion. For a derivation of the bending energy of a solid film in terms of its compressional and shear elastic constants, see Ref. 4.

Explain why the curvature elastic constants vanish in the limit of vanishing shear modulus. Include in your explanation a consideration of why one expects k and $\bar{k}$ in a small molecule liquid (*i.e.*, with no springlike chain molecules such as those considered in the text) to be very small.

In Ref. 4, it is shown that the bending moduli scale with the cube of the thickness of the solid thin film. A similar result can be obtained from the general discussion in this chapter, since the modulii scale with the *integral* of the second moment of the transverse pressure profile. However, in a solid, thin film, the possibility for incoherent bending of the layers also exists — *i.e.,* where the strain is not continuous throughout the thickness of the film. In this case, the bending modulus is expected to be linear in the layer thickness. If the solid bends incoherently, what energy cost has to be paid? What are the limiting conditions when you might expect coherent and incoherent bending of a solid, thin film?

What physical properties would generate a spontaneous curvature in a solid film?

2. Bending of Anisotropic Solid Films

Find the curvature modulii, k and $\bar{k}$, of an anisotropic solid film composed of "layers" where the in-plane symmetry is hexagonal (use the form of the compressional and shear energy expansion for anisotropic media in Ref. 4), in the limit that the in-plane shear elastic constant vanishes. Assume that the film and its elastic constants are uniform throughout its thickness. Compare this case with the isotropic solid film in the limit of vanishing shear modulus.

3. Neutral Surface

Based on the discussion in the text, calculate the bending modulus, g_2, for deformations relative to the neutral surface in terms of the parameters that describe the bending at some other surface (*e.g.,* the polar-nonpolar interface), f_1, f_1', f_0''.

4. Bilayer Curvature Energy

Derive an expression for the curvature elastic free energy of a bilayer in terms of the curvature elastic constants of the monolayer. Treat the case where the two monolayers are equivalent and noninterpenetrating, so that one adds their curvature energies, but note that the monolayers each have a finite thickness, which makes their curvatures inequivalent. Compare with the Helfrich form and comment on the effective saddle curvature as a function of the spontaneous curvature of each monolayer.

5. Bending of a Charged Membrane

Consider a single, charged membrane with a surface charge per unit area, σ, in the presence of a large concentration of electrolyte so that the potential gradients are always small and the electrostatics can be solved in the Debye-Hückel approximation (see the problems at the end of Chapter 5). Calculate the product $k\,c_0$ and saddle-splay curvature elastic constant for this system using

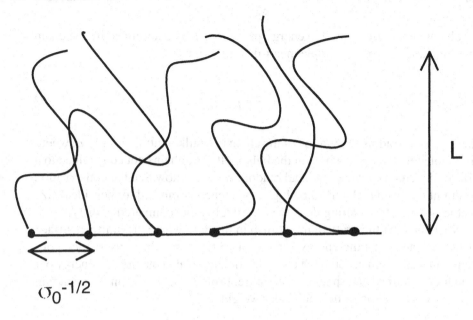

$\sigma_0^{-1/2}$

Figure 6.6 Polymers grafted to a surface with an average distance between chains proportional to $\sigma_0^{-1/2}$. The height of the brush is L.

the equations for the pressure distribution. Comment on the scaling of these quantities with the screening length.

6. Persistence Length of a Polymer

Find the tangent-tangent correlation function and persistence length for a polymeric chain with curvature elasticity — *i.e.*, a one-dimensional membrane embedded in a three-dimensional space. This is applicable to the physics of a flexible rod undergoing thermal fluctuations. Contrast the results with those of a two-dimensional membrane.

7. Curvature Elasticity of Polymer Brush

Consider a set of polymeric chains chemically grafted to a surface with an average area density of σ_0 (grafted polymer brush shown in Fig. 6.6) in the case where there is no solvent penetration into the chains (dry or melt brush). The brush is described by a brush height, L, which is related to σ_0 and the degree of polymerization of each chain, N, by the constraint that the density of monomers is constant. This yields a conservation law relating L, σ_0, N. For this case of no solvent penetration, the free energy per chain, f, is determined by the chain stretching and is given in simple mean-field theory by

$$f = \frac{3T}{2} \frac{L^2}{L_0^2} \tag{6.64}$$

where L_0 is the radius of gyration of the chain in a bulk melt: $L_0 = N^{1/2}a$, where a is a molecular size and where in the bulk melt (*i.e.*, chains not constrained to a surface) the chain executes an ideal random walk[12]. Show that the conservation constraint causes the chains to be highly stretched compared to their ideal size — at least when the grafting density is relatively large compared with $1/N$.

Now consider the grafted, melt brush in the case where the grafting surface is curved. The stretching energy has the same form, but the conservation constraint must take into account the curved geometry. Calculate the free energy per chain for cylindrical and spherical curvature. From these results find the bending modulus as a function of the molecular weight.

8. Unbinding of Membranes

Consider a stack of membranes with an average spacing d, with only entropic (Helfrich) interactions. What will happen if more solvent is added to the system? Now, consider the same stack, but with an additional attractive interaction between the membranes. What do you expect to happen if the attraction is strong enough and how might this affect what happens if one now tries to dilute the membranes by adding more solvent. (See Ref. 13 for a review.)

9. Membrane Between Walls

Consider a membrane whose physics is described by a Hamiltonian consisting of the curvature energy with *zero* spontaneous curvature, which is constrained to lie between two hard walls separated by a distance D. Describe the conformation of the membrane by its height, $h(x, y)$, relative to the plane at $z = 0$, located at the midplane between the two walls. Assume that the effect of the walls adds to the Hamiltonian an effective harmonic potential of the form: $\frac{1}{2}\gamma h^2$ which acts on the membrane in addition to the bending energy (assume zero spontaneous curvature and that the saddle-splay energy can be neglected).

Determine γ so that the mean square fluctuation of the membrane at any point in the xy plane is equal to μD^2 where μ is a number of order unity that is assumed to be known. What is γ for $D \to \infty$?

Use this Hamiltonian to calculate the free energy of the system as a function of the spacing between the walls. What is the difference in the free energy between the membrane with walls and a free membrane? Is it higher or lower than the free energy of the isolated membrane and why?

Could this result have been predicted from dimensional analysis and why or why not?

6.8 REFERENCES

1. See, for example, the articles by Y. Kantor and D. Nelson, *Statistical Mechanics of Membranes and Surfaces*, eds. D. Nelson, T. Piran, and S. Weinberg (World Scientific, Teaneck, NJ, 1989) pp. 115 and 137 respectively. Simulations of self-avoiding surfaces are discussed in F. F. Abraham and D. R. Nelson, *Science* **249**, 393 (1990) and *J. Phys. (France)* **51**, 2653 (1990).

2. Z. G. Wang and S. A. Safran, *J. Phys. (France)* **51**, 185 (1990).

3. W. Helfrich, *Z. Naturforsch.* **28c**, 693 (1973).

4. L. D. Landau and E. M. Lifshitz, *Theory of Elasticity*, 2nd ed., revised and enlarged (Pergamon, New York, 1970).

5. W. Helfrich in *Physics of Defects*, Les Houches, Section XXXV, eds. R. Balian *et al.* (North Holland, Amsterdam, 1981).

6. I. Szleifer *et al.*, *J. Chem. Phys.* **92**, 6800 (1990).

7. P. G. Higgs and J. F. Joanny, *J. Phys. (France)* **51**, 2307 (1990).

8. P. G. de Gennes and C. Taupin, *J. Phys. Chem.* **86**, 2294 (1982).

9. W. Helfrich, *J. Phys. (France)* **46**, 1263 (1985).

10. L. Peliti and S. Leibler, *Phys. Rev. Lett.* **54**, 1690 (1985).

11. W. Helfrich, *Z. Naturforsch.* **33a**, 305 (1978).

12. P. G. de Gennes, *Scaling Concepts in Polymer Physics* (Cornell University Press, Ithaca, NY, 1979).

13. R. Lipowsky, *Nature* **349**, 475 (1991).

CHAPTER 7

Colloidal Dispersions

7.1 INTRODUCTION

The previous two chapters focused on the properties of and interactions between *flat* interfaces or nearly flat membranes. In these systems, the essential degree of freedom is related to the spacing between the interfaces or membranes, and we discussed how the interaction energies or free energies depend on this distance. Here, we discuss the statistical thermodynamics of colloidal dispersions that consist of solid or fluid "particles" in a solvent. The size of these particles is typically much greater than the size of the solvent molecules, thus differentiating these dispersions from molecular solutions where all the components have roughly the same size scale. In colloidal dispersions there are two important lengths: (i) the distance between colloidal particles and (ii) the size of the particles. We focus on the problem of colloid stability and how electrostatic and steric interactions stabilize such dispersions. The chapter begins with a description of colloids and their uses and continues with a discussion of the effects of attractions that lead to colloidal aggregation and destabilization. A discussion of phase separation due to attractive interactions between hard spheres is presented. It is shown that the dependence of the interaction energy on the sphere size is an important quantity that determines the stability of colloidal solutions.

Since surface interactions are most easily calculated in the flat plate geometry, we present the Derejaguin approximation which relates the interactions between flat and spherical interfaces. We then discuss the so-called DLVO theory of colloid stability via electrostatic interactions and outline how polymeric coatings are an additional source of repulsion via a discussion of "polymer brushes". Although most of the chapter focuses on the equilibrium properties of colloidal dispersions, it is of interest to consider the kinetics of the structures of colloidal aggregates when the dispersion is unstable and in this connection we discuss the interesting (nonequilibrium) phenomenon of the fractal structures of colloidal aggregates.

7.2   COLLOIDAL DISPERSIONS

Types of Colloidal Dispersions

While many of the simple, physical properties of interfaces and membranes can be studied by focusing on either single, or one-dimensional stacks of nearly flat systems, there is considerable interest in systems with a macroscopic amount of interface composed of finite-sized "particles": **colloids**. While a one-dimensional stack of membranes is a particular type of colloid already discussed at the end of Chapter 6, we focus here on isotropic systems composed of spherical or cylindrical particles dispersed in a fluid. When these particles are much larger than a molecular size, their interactions can be related to their interfacial properties. A dispersion of one species in another where the size scale that characterizes the dispersed phase is a molecular length, is well described as a molecular solution; what is interesting about a colloid is its supermolecular nature. Interactions between flat interfaces, discussed in Chapter 5, provide a starting point for understanding such systems, with several modifications as discussed below. In this section, we discuss nondeformable systems — the particles maintain their integrity and the focus is on their mutual interactions, which lead to correlations and possibly to phase separations. An important class of deformable colloidal dispersions are polymers in solution; the macromolecules change their conformations as a function of the solvent quality and/or polymer concentration. In the next chapter, we discuss self-assembling colloids such as micelles, vesicles, and microemulsions, where even the number of molecules that compose a "particle" (as well as its size and shape) can vary as a function of concentration, chemical conditions, or temperature.

Applications of Colloids

Colloidal systems exist in both nature and industry and can consist of either solids or liquids dispersed in either fluids or gases. Blood is a dispersion of the red blood cells (which are similar to self-assembling colloids) in serum and emulsions or microemulsions (see Chapter 8) are dispersions of oil in water or water in oil. Fog, mist, and smoke are dispersions of small particles in gases, while pollution control deals with dispersions of solid particles in air. Foams (dispersion of liquid in a gas at relatively high volume fractions of liquid) are familiar from toothpastes to beer. Many industrial processes make use of colloidal dispersions of solid particles in fluids to tailor the hydrodynamic properties of the fluid or sometimes to produce a system with large amount of internal surface area for catalytic applications.

Colloid Stability

The most striking and useful property of colloidal dispersions such as gold or polystyrene spheres in water or silica particles in oils, is their stability. Namely, in a certain range of volume fractions, ϕ, the particles remain in solution. As the system is made more concentrated, as evidenced by increased milkiness (due to an increase in the scattering of light), the particles may tend to aggregate. The process of aggregation is called **flocculation**; a dense phase of particles tends to separate from the solvent into a solidlike precipitate[1]. For uncharged systems with strong attractive interactions between the particles (*e.g.*, due to dispersion forces) the aggregates can be a "hard sphere" solid or glass with the average interparticle distance close to that of the particle diameter. If the attraction is extremely "sticky", the nonequilibrium structure of the aggregate may be dominated by diffusion effects (diffusion-limited aggregates) and may be fractal in nature[2,3]; the particles tend to stick to the outside of growing clusters and never fill in the "holes". These fractal clusters can have very low densities and hence remain in solution at relatively high volume fractions. On the other hand, in systems where the interparticle repulsions dominate the attractions, the particles may form *colloidal crystals*[4]. In particular, if the repulsion is relatively long-range these crystals can have lattice spacings that are much larger than the actual particle diameter. The order-disorder transition in these systems can therefore occur at relatively small volume fractions; in pure hard-sphere systems, the order-disorder transition occurs at a volume fraction[5] of $\phi = 50\%$. Since colloidal particles can range in size from tens of Angstroms to microns, the lattice spacings of some colloidal crystals can be comparable to the wavelength of light. The resulting Bragg reflections from these crystals occur in the visible region and can be seen by eye as iridescent bands in the test tube. The crystal is very "soft" since the interparticle spacings are so large and small shears (gentle shaking) will disrupt the long-range order and the resulting iridescence.

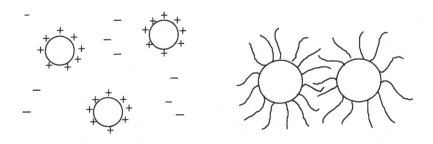

Figure 7.1 Colloidal dispersions stabilized by electrostatic repulsions (left) and steric repulsion by surface grafted polymers (right).

Stabilizing Colloids with Repulsive Interactions

The fundamental property of interest in colloidal dispersions is their stability against flocculation. Since van der Waals forces tend to produce attractions between like particles in an unlike fluid dispersant, one would expect most colloidal dispersions of large enough particles to flocculate. This is because the attraction per particle increases with the particle size; when the net attraction per particle is much larger the entropy of dispersion (*i.e.*, the translational entropy, which is roughly speaking $k_B T \log \phi$ per particle), the system is expected to phase separate with an equilibrium between a phase with high density of colloidal particles coexisting with a phase with a low density of colloid. Indeed, this is most often the case if the particle surface is untreated. For many colloidal dispersions, the attractions due to van der Waals forces are much greater than the dispersion entropy of $k_B T \log \phi$ per particle, and to stabilize such dispersions at reasonable concentrations it is necessary to treat the particle surface so that the particles interact repulsively. This is done (see Fig. 7.1) by introducing charged species on the surface (*e.g.*, ionic surfactants) whose counterions are soluble in the solvent. In this case, the electrostatic repulsions at long distances can produce large energy barriers that prevent aggregation. Another way to introduce repulsions is by grafting polymers onto the particles; the polymer layers "bump" up against each other when two particles come near and because of the entropy loss of the chains, resist compression. They thus prevent aggregation. A quantitative description of these effects is presented below; we first present a general picture for the understanding of the thermodynamic stability of interacting colloids.

7.3   DISPERSIONS OF INTERACTING PARTICLES

Equilibrium Behavior

While colloid stability is generally discussed in terms of nonequilibrium, irreversible flocculation, much of the underlying physics can be understood by focusing on the phase behavior of equilibrium ensembles of interacting particles. These systems can be understood from equilibrium considerations if the strength of the attractive interactions is comparable to $k_B T$; if it is much larger, the time for two particles to come apart after contact may be much longer than the experimental time scale and the system cannot be thought of as being in equilibrium. For repulsive interactions, this consideration does not apply and a description in terms of thermal equilibrium is more generally justified.

Hard-Sphere Suspensions

The simplest colloidal dispersion is one of hard spheres, which interact with a potential

$$V_d(\vec{r}) = \infty \qquad (7.1a)$$

for $r < d$ and

$$V_d(\vec{r}) = 0 \qquad (7.1b)$$

for $r > d$, where d is the particle diameter. The properties of such systems have been studied as models for systems of rare-gas atoms and a host of approximation schemes for treating the statistical mechanics of such systems have been developed[5]. At low volume fractions, ϕ, a virial expansion for the free energy per particle, f_s (see Chapter 1), can be used:

$$f_s = T \left[\phi(\log n v_0 - 1) + 4\phi^2 + 5\phi^3 + v_4 \phi^4 + v_5 \phi^5 \right] \qquad (7.2)$$

where the fourth and fifth virial coefficients have values[5] of $v_4 = 6.12$ and $v_5 = 7.06$, respectively. The concentration, n is related to the particle diameter and volume fraction via $n = \pi d^3 \phi/6$ and v_0 is an atomic volume. At higher volume fractions, an accurate description of the system is given by molecular dynamics and Monte Carlo simulations, which predict that the osmotic pressure in the fluid phase, $\Pi = -\partial F/\partial V$ (where F is the total free energy and V is the volume) diverges at a volume fraction of $\phi \approx 0.64$, which is the volume fraction for random close packing. However, this divergence is preempted by a first-order transition from a disordered fluid to a face-centered-cubic solid for $\phi = 0.50$; in the region from $0.50 < \phi < 0.55$ the fluid and solid phases coexist. Thus, at high enough volume fractions, colloidal crystals exist even in hard sphere systems. In charged systems, the crystalization can take place at much smaller volume fractions, as low as $\phi \approx 10^{-3}$ due to the presence of long-range repulsive interactions.

Attractive Interactions

Within the virial expansion, the presence of longer range repulsions and/or attractions can be accounted for by adding to the free energy of Eq. (7.2) a term

$$f_a = -\frac{T}{2}\phi^2 b(d, T) \tag{7.3}$$

$$b(d, T) = n_0 \int d\vec{r} \left[e^{-\tilde{V}(\vec{r})/T} - 1 \right] \tag{7.4}$$

where $n_0 = (\pi d^3/6)^{-1}$ is the inverse of the volume per sphere and where d is the particle diameter that enters into the interaction potential, $\tilde{V}$, which includes all interactions except the hard core. The integral in Eq. (7.4) is over all interparticle separations, $\vec{r}$, permitted by the hard-core repulsion. In general, the attractive interactions which enter into $\tilde{V} < 0$ also result in higher order terms in ϕ. However, if the range of the attractive potential is small compared with the particle radius (or equivalently, the attraction strength is much less than $k_B T$ even when the two particles are separated by a distance small compared with d), these higher order terms may be neglected compared with the contributions from the hard sphere terms. To order ϕ^3, we write the total free energy per particle, $f = f_s + f_a$, as

$$f = T \left[\phi(\log n v_0 - 1) + \tfrac{1}{2}a\phi^2 + 5\phi^3 \right] \tag{7.5}$$

where the virial coefficient:

$$a(d, T) = 8 - b(d, T) \tag{7.6}$$

is positive for small attractive interactions and negative for large attractions.

It is important to note that the dependence of the virial coefficient on the particle diameter comes from the dependence of the interaction $\tilde{V}$ on d. Generally speaking, one imagines, that the larger the particle size, the greater the number of "contacts" between the molecules on two adjacent particles. Thus, for example, for attractive interactions (where $\tilde{V} < 0$ and $b > 0$) one expects $\tilde{V}$ and hence the term, $b(d, T)$ to be an increasing function of d. In contrast to this dependence of the attractive interaction on d, the contribution to the total virial coefficient, $a(d, T)$, due to the hard sphere interactions (the coefficient of ϕ^2 in Eq. (7.2), which results in the factor of 8 in Eq. (7.6)) shows no dependence on the particle size. This is because the entropy loss due to excluded volume is $k_B T$ per particle; the attractive interaction *per particle* as expressed by $b(d, T)$ does increase with the particle size. It is the competition of these two effects that determined the overall sign of $a(d, T)$ and hence the tendency to phase separate or remain in a single, homogeneous phase. One expects, therefore, that as the size of the particles is increased there comes a point when $a(d, T)$ becomes sufficiently negative and phase separation (flocculation) occurs. Thus, large enough colloidal particles should always exhibit flocculation since at some particle size, the van der Waals attraction between them must overcome the excluded volume repulsions.

It is therefore of interest to estimate the dependence of the attractive interaction, $b(d, T)$ on the sphere size. For relatively short-range interactions between spheres, the attractive interaction per sphere can be approximately obtained from the interaction between two flat plates. As we now show, this approximation results in an effective attraction between spheres that increases *linearly* with the sphere radius. Once the interaction energy per sphere is known, the virial coefficient can be calculated from Eq. (7.4) and the stability of the system to phase separation is related to the magnitude of this coefficient. Thus, the general plan is to (i) find the effective interaction between two flat plates, (ii) relate this interaction energy to an interaction between spheres via the Derejaguin approximation, (iii) calculate the virial coefficient for the spheres in order to assess the stability of the single-phase colloidal dispersion.

Derejaguin Approximation for Spheres

While the interactions based on flat plate energies are a useful guide for understanding colloid stability, it is also of interest to know the interaction energy per sphere — since it is this energy that is important in determining colloid stability with respect to an entropy of $k_B T$ per sphere. For long-range interactions, where the length scales of interest are much bigger than the sphere size, a point-particle approximation can be used. For shorter range interactions, the geometrical effects yield a dependence of the interaction energy on the sphere size in a general manner first proposed by Derejaguin.

The effective interaction is calculated by noting that when the distance between the spheres, D, is smaller than the range of the interaction (see Fig. 7.2 for a description of the geometry involved), one can evaluate the interaction potential by integrating the potential for flat plates over the range of distances in the gap between the spheres. In cylindrical polar coordinates (with the coordinate r orthogonal to z) the surface to surface separation in the gap is

$$z(r) = D + 2R \left[1 - \sqrt{1 - \frac{r^2}{R^2}} \right] \approx D + r^2 \frac{1}{R} + ... \qquad (7.7)$$

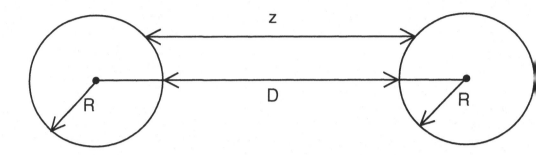

Figure 7.2 Geometry for the Derejaguin approximation.

where the terms omitted are of order D/R where R is the sphere size. Thus,

$$dz = \frac{2r}{R}dr \qquad (7.8)$$

and the total effective interaction, V_e, between the lamellae of area $2\pi rdr$ is obtained by integrating the interaction energy per unit area over the distribution of areas of each "disk"-like section of the spheres. We then find

$$V_e = 2\pi \int_0^\infty W(z)\,rdr \qquad (7.9)$$

Here, W is the interaction energy *per unit area* of two flat plates as a function of their spacing, which is locally given by z. Using Eq. (7.8) in Eq. (7.9) we find

$$V_e = \pi R \int_D^\infty W(z)\,dz \qquad (7.10)$$

The interaction energy per sphere is *linear* in the sphere radius. If the flat plate interaction W has a characteristic range λ, the interaction energy per sphere will scale like λRW. This allows predictions as a function of sphere size. Furthermore, Eq. (7.10) allows a simple estimate of the force, $\vec{F}$, between two spheres whose magnitude is given by

$$F = -\frac{\partial V_e}{\partial D} = \pi RW(D) \qquad (7.11)$$

and whose direction is along a line joining the centers of the two spheres. The approximation, therefore allows us to predict the dependence of the interaction potential on the sphere size and separation; it is this effective potential that is used in Eq. (7.4) to compute the virial coefficient.

Liquid-Gas Phase Separation

The free energy of the hard-core system, Eq. (7.2) augmented by a short-range attractive term, Eqs. (7.3,7.4), $f = f_s + f_a$, shows an instability to phase separation into coexisting low-density (gas) and high-density (liquid) phases. To determine the coexistence curve (*i.e.*, the locus of points where phases with the densities ϕ_1 and ϕ_2 are in equilibrium), one must solve for the equilibrium conditions of equal chemical potentials

$$\mu = \left(\frac{\partial f}{\partial \phi}\right)_1 = \left(\frac{\partial f}{\partial \phi}\right)_2 \tag{7.12}$$

and equal osmotic pressures

$$f(\phi_1) - f'(\phi_1)\phi_1 = f(\phi_2) - f'(\phi_2)\phi_2 \tag{7.13}$$

where $f' = \partial f/\partial \phi$. Within this mean-field approach, one can obtain the critical values of the virial coefficient, b_c, and the volume fraction, ϕ_c, at which phase separation will first occur, by examining Eqs. (7.12,7.13) for the case that $\phi_1 \rightarrow \phi_2$ (*i.e.*, where the distinction between the two phases vanishes). One finds that ϕ_c and b_c are determined by

$$\frac{\partial^2 f}{\partial \phi^2} = 0 \tag{7.14a}$$

$$\frac{\partial^3 f}{\partial \phi^3} = 0 \tag{7.14b}$$

Using the virial expansion described previously (with terms up to the fifth virial coefficient for the hard sphere part of the potential and only the second virial coefficient for the attractive part), we find that $b_c \approx 21$ and $\phi_c \approx 13\%$. The existence of this phase separation is an indicator of flocculation and that the colloid will be unstable and will not remain dispersed in solution in this two-phase region.

Hard Spheres Plus Longer Range Repulsions

A system of spheres can often be described by a hard-core potential, with a diameter d_0, plus a longer range, "softer" repulsive part, $V_r(\vec{r})$. Often these softer potentials can be treated by various perturbation schemes as detailed in Chapter 6 of Ref. 5. One such scheme is to utilize the variational approach discussed in Chapter 2 with a hard-sphere system with an *effective* diameter, d chosen as the variational parameter.

Using the variational principle described in Chapter 2, the approximate free energy, F, is given by

$$F = F_d + \langle V(\vec{r}) - V_d(\vec{r}) \rangle_d \qquad (7.15)$$

where the subscript d refers to an effective hard-sphere system with particles of diameter, d, whose potential is $V_d(\vec{r})$ and whose properties are known in a variety of approximation schemes. The actual repulsive potential is $V(\vec{r})$ which may have a hard core for $r < d_0$, and we always choose $d_0 < d$ — i.e., the effective hard-core diameter is always larger than the actual hard-core diameter for purely repulsive potentials. The perturbation term involves the thermal average of the potentials with the probability distribution of the hard-sphere model system. Thus, if we denote by $g_d(\vec{r})$ the pair-correlation function (the probability that there is a particle at a distance $\vec{r}$, given that there was a particle at $\vec{r} = 0$), for the hard-sphere system with diameter d,

$$\langle V(\vec{r}) - V_d(\vec{r}) \rangle_d = \int d\vec{r}\, g_d(\vec{r})\, \left[V(\vec{r}) - V_d(\vec{r}) \right] \qquad (7.16)$$

Although one might question the applicability of this kind of perturbation theory to hard-core interactions (where the potentials become infinite), we note here that the term $g_d(\vec{r}) \left[V(\vec{r}) - V_d(\vec{r}) \right]$ never diverges since (i) $d_0 < d$ so divergences due to the actual hard core in V are subtracted off by the hard core of the effective potential, V_d, and (ii) the pair-correlation function, g_d vanishes when $r < d$.

The free energy, F, is then minimized with respect to d to find the hard-sphere diameter, d, that best describes the actual system. To first order, all of the properties of the actual system are given by the corresponding hard-sphere properties, with an effective diameter, d. A detailed comparison of different approximation schemes, all of which are perturbations about the hard-sphere potential, is discussed in Ref. 5.

This procedure defines an *effective* volume fraction, $\phi^* = \phi(d/d_0)^3$, where d_0 is the hard-core diameter of the actual system. The softer part of the (repulsive) potential defines the effective diameter $d > d_0$ so that the effective volume fraction can be considerably greater than the actual volume fraction, ϕ. Thus, transitions such as crystalization will occur when $\phi^* = 0.50$; the actual volume fraction, ϕ, can be substantially less than this value if there is an enhancement of d by the soft-core repulsions; this enhancement is particularly effective since the effective volume fraction scales as $(d/d_0)^3$.

7.4 COLLOID INTERACTIONS: DLVO THEORY

Balance of Interactions

The standard theory of colloid aggregation is referred to as the **DLVO theory** (Derejaguin-Landau-Verwey-Overbeek) and describes the balance between the van der Waals attractions and electrostatic repulsions of colloidal particles. At high salt concentrations, when the electrostatics is sufficiently screened, the attractive dispersion forces result in flocculation. For equilibrium systems, one can analyze the balance of interactions and find the critical virial coefficient, b_c, described earlier. However, when there exists a strong, short-range attraction (due to dispersion forces), and a longer range repulsion (due to the screened electrostatic interactions), the aggregation may become nonequilibrium. While it may be true that the global minimum free energy state is the completely aggregated colloid, there may exist secondary minima in the free energy (see Fig. 7.3) as a function of the interparticle spacing which can give rise to very long-lived metastable states where the colloid is unaggregated (*i.e.*, with the particles at finite distances from each other). These states can be separated from the aggregation minimum by an energy barrier determined by the maximum height of the total potential energy. If that barrier is much larger than $k_B T$, it will not be overcome by the Brownian motion of the particles; they will never feel the extremely strong van der Waals attraction at close range. Flocculation is prevented in this case. As more salt is added to the system, the range of the repulsive interaction decreases and the system flocculates at relatively small volume fractions.

Interactions of Flat Plates

We first consider the balance of interactions for flat plates. Once this interaction is understood, the interaction between two spheres can be obtained within the Derejaguin approximation; the virial coefficient and hence the stability of the system to phase separation can then be determined.

From Chapter 5, the attractive dispersion forces for two plates separated by a gap D, gives rise to an energy per unit area:

$$v_a = -\frac{A}{12\pi D^2} \tag{7.17}$$

where A is the Hamaker coefficient. For the case of short-range repulsions (added salt) the screened Coulomb repulsion between the plates results in an energy per

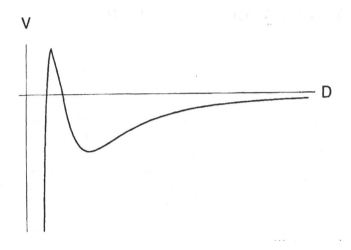

Figure 7.3 Schematic representation of the combined van der Waals attraction and electrostatic repulsion — the DLVO potential with a steep attractive well at small separations, D, and a barrier to reaching this attractive well. There is also a secondary attractive minimum.

unit area as a function of distance D, that has the form

$$v_r = \frac{\beta T n_0}{\kappa} e^{-\kappa D} \qquad (7.18)$$

where the screening length, κ^{-1} obeys: $\kappa^2 = 8\pi n_s \ell$, where n_s is the bulk salt concentration and $\ell = e^2/(\epsilon T)$ is the Bjerrum length (see Chapter 5) and β is dimensionless and independent of D. This expression comes from a linearization of the Poisson-Boltzmann equation for case where the salt is in equilibrium with a bulk reservoir. In this limit, the two interfaces interact via short-range repulsions since at high salt concentrations, $\kappa D \gg 1$. (In the limit of no added salt, the interaction energy is a power law that can more easily dominate the van der Waals attractions.) The total potential energy $V = v_a + v_r$ is shown in Fig. 7.3. It is characterized by a strong minimum at small distances, a repulsive barrier, and a shallow minimum at larger distances; this latter minimum can give rise to "gas-liquid" transitions as discussed earlier, while the strong minimum will give rise to flocculation if the repulsive barrier is not sufficient to prevent the particles from close approach.

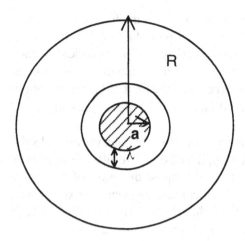

Figure 7.4 Geometry of charge distribution used in the variational approximation for the spherical geometry.

7.5 LONG-RANGE ELECTROSTATIC INTERACTIONS

Charged Colloids

Experiments on charged colloids typically deal with particles with dimensions on order of a micron with 100 to 1000 charged groups. If there is very little added salt (low ionic strength), the electrostatic repulsions are strong enough to stabilize ordered structures. These systems can function as classical models of solids. Many experiments that probe the crystalinity of these systems are possible, such as melting as a function of ionic strength, structural (fcc-bcc) phase transitions, and shear melting. We shall consider the interactions of spherical colloidal particles in the limit of no added salt, since it is in this regime that the most dramatic effects of long-range interactions are observed. It is in this case that the colloidal crystals are stabilized at very small volume fractions of the colloidal particles due to the long-range repulsions. For large amounts of added salt (compared to the counterion density), the system interacts via a screened Coulomb interaction as described previously.

Counterion Charge Density

To analyze the electrostatic interactions of even a periodic array of charged spheres, one should solve the Poisson-Boltzmann equation within a primitive unit cell of the system, which reflects the periodicity. At the boundaries of this "Wigner-Seitz" cell, there is no electric field, since by symmetry, the cell encloses a volume where the counterion charge density balances the fixed charge. There are two difficulties with this approach: (1) the Wigner-Seitz cell for a typical colloidal crystal has a complicated geometry and (2) even for a spherical Wigner-Seitz cell, there is no analytic solution of the Poisson-Boltzmann equation in the no-salt limit, for the case of spherical symmetry. In the limit of large separations between the particles (of interest in the dilute crystals), the approximation of nearly flat interfaces fails.

However, much of the physics[6,7] can be understood using a variational solution of the free energy for the charge density. Based on the results of the array of flat, charged layers discussed in Chapter 5, we suppose that most of the charge is localized in a region very close to the spherical surface, near the fixed charge, while the rest (a small fraction of the total charge) is distributed uniformly in space. From the one-dimensional solution (appropriate for distances close to the sphere's surface), the characteristic distance over which most of the charge is localized is $\lambda = 1/(\pi\sigma_0\ell)$ (see Chapter 5), where σ_0 is the charge density and $\ell = e^2/(\epsilon T)$. We assume that the particle radius $a \gg \lambda$. As shown in Fig. 7.4, we take as the variational charge distribution within each spherical Wigner-Seitz cell of radius R a charge $(Z - Z^*)$ uniformly distributed in an inner region $a < r < a + \lambda$, and an *additional* charge Z^* uniformly distributed in the *entire* region $a < r < R$. We thus write:

$$n_i(r) = \frac{(Z - Z^*)}{V_i} + \frac{Z^*}{V} \tag{7.19}$$

for the inner region $a < r < a + \lambda$ with volume

$$V_i = \frac{4\pi}{3}\left[(a + \lambda)^3 - a^3\right] \tag{7.20}$$

The outer volume is

$$V = \frac{4\pi}{3}\left[R^3 - a^3\right] \tag{7.21}$$

In the outer region, $(a + \lambda) < r < R$, the charge density is written:

$$n_0(r) = \frac{Z^*}{V} \tag{7.22}$$

The variational parameter is the effective[6] free charge, Z^*; in a more refined treatment, the distance λ can also be varied and the assumption of uniform density can be relaxed.

Free Energy Variation

We consider the limit of a dilute dispersion of spheres so that $R \gg a$. In addition, we consider the high charge density limit, $\lambda \ll a$. Using the variational ansatz[7] for the charge density, the free energy per charge, F/Z (see Chapter 5) is written in these limits as

$$f = f_0 + T\alpha \left[(1 - \beta)^2 - 2(1 - \beta)\right] + T\beta \log[\phi\beta]$$

$$+ T(1 - \beta) \log \left[\frac{(1 - \beta)a}{3\lambda}\right] \tag{7.23}$$

In Eq. (7.23), $\alpha = 2\pi\sigma_0 \ell a$. Here the variational parameter is $\beta = Z^*/Z$, which is assumed to be small; ϕ is the volume fraction, which is related to R by $\phi = (a/R)^3$. The first term in Eq. (7.23) is a constant, independent of β. The next term represents the Coulomb repulsion of the counterions in the inner region, while the term proportional to $(1 - \beta)$ is their attraction to the fixed charge. The last two terms are the entropies of the "free" (unbound) charge in the outer region and of the charge in the inner region respectively. The terms representing the Coulomb energies associated with the unbound charge in the outer region give higher order terms in the volume fraction ϕ and are neglected in the limit of $R \gg a$. Variation of the free energy with respect to $\beta = Z^*/Z \ll 1$ yields

$$\beta = \frac{Z^*}{Z} \approx -\left(\frac{1}{2\alpha}\right) \log \left[\frac{3\lambda\phi\beta}{a}\right] \tag{7.24}$$

The fraction of *delocalized* or free charge is approximately a constant with a logarithmic *increase* as the volume fraction of spheres is decreased. This occurs because the entropy of the counterions becomes more important for small ϕ, thus stabilizing the free charge. The fraction of charge that is free or unbound is much larger than that of the one-dimensional case where the entropy is much more restricted and either another variational calculation or the exact result (Eq. (5.72)) implies β *decreases* linearly with ϕ: $\beta = (Z^*/Z) = 1/(\ell\sigma_0 D) \sim \phi$. This qualitative difference in the behavior of the charge renormalization is due to the increased entropy in the spherical case. As discussed in Chapter 5, the force between two *flat* surfaces is simply related to the charge density at the symmetric boundary located at $z = 0$:

$$\Pi_{zz} = Tn(0)$$

This relation does not hold exactly for an array of spheres. However, heuristically, one might expect that for dilute arrays, the force per unit area between two of the spheres may be related to $Tn(\vec{r}_b)$ where $\vec{r}_b$ represents the Wigner-Seitz cell boundary. This simple form for the pressure arises because the electric field vanishes on the boundary; one is thus left only with the osmotic pressure of the counterions. From the previous discussion, $n(R) \sim Z^*/R^3 \sim \beta Z\phi/a^3$ and can still be sufficient to stabilize the crystal, even though the effective charge may be small compared to Z.

7.6 STERIC INTERACTIONS: POLYMER ADSORPTION

Steric Repulsion

In addition to electrostatic stabilization of colloids, the steric repulsion of attached polymers or surfactants is another method used to keep colloidal particles from aggregating. While an equilibrium solution of such solutes that interact with surfaces usually results in an effective attractive interaction between these walls (see Chapter 5), nonequilibrium systems can show repulsive interactions. This is because in the equilibrium case, where there is a reservoir of solution in equilibrium with a solution in between the walls, the concentration of solute can adjust as the distance between the walls is changed. If the distance between the walls becomes smaller than the correlation length (or molecular size, radius of gyration) of the solute, there can be a depletion of solute in the region near or at the walls; the solute just escapes to the reservoir. However when these molecules are attached to the walls with strong chemical bonds (the time scale for desorption being much longer than the time scale of the observations), the solute molecules cannot escape from the region near the walls and the effective interaction between the walls can be repulsive. This effect can occur either for the case where (i) all the monomers that compose the polymeric molecule are attracted to the wall (polymer adsorption)[8], or (ii) one end of the polymer is chemically modified so that it is strongly attracted to the wall (grafted polymer chains). In the case of adsorption, the binding energy per polymeric chain can be many times $k_B T$ even though the individual attraction energies of each monomer to the wall are small, because *all* of the monomers are attracted to the wall. In the case of the grafted chains, there is usually a strong chemical interaction between the end-monomer and the wall to maintain the grafted state. In what follows, we treat the grafted case; the adsorbed case is discussed in Refs. 8,9.

In the case of chains grafted to a surface, the repulsive force between two such treated surfaces is related to the changes in density and in the polymer stretching caused by compressing the polymer layers on each colloidal particle. For simplicity, we consider long-chain polymers attached to flat surfaces at fixed positions (so-called grafted polymer brushes) and estimate the repulsive interaction between these brushes from a calculation of the energy required to compress a brush to a height that is smaller than its equilibrium height. This

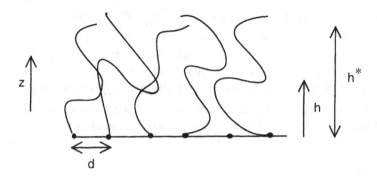

Figure 7.5 Polymers grafted to a surface with an average spacing, d, and height, h^*.

compression would arise from the presence of a neighboring colloidal particle, which is assumed to act as a solid barrier; interdigitation of the polymers from the two neighboring surfaces is neglected.

As before, we consider the interaction of two flat plates coated by grafted polymeric layers. Once the interaction between the two flat surfaces is known, the Derejaguin approximation and virial expansion can be used to determine the stability of the spherical colloidal dispersion as described previously.

Polymer Brush

The physics of the polymer brush (see Fig. 7.5) has been treated in several approximations. In the simplest approach, the polymers are all assumed[10] to terminate on the same surface $z = h^*$, while a more accurate description takes into account the distribution of ends in the region between the grafting surface and the top of the brush in a self-consistent manner[11]. For simplicity, we consider an intermediate approach[12] where the probability to find a polymer end is proportional to the density, $c(z)$. We consider a polymer in a **good solvent** which is defined by a positive second virial coefficient characterizing the monomer-monomer interactions (the term proportional to c^2 in the free energy). A **poor solvent** is characterized by a negative second virial coefficient that tends to cause the chain to collapse (*i.e.*, to prefer a locally large value of c). This effect is stabilized by the higher order, positive, virial coefficients. A **theta solvent** is characterized by a vanishing second-order virial coefficient.

Thus, for a good solvent, the free energy describing the polymer layer has a contribution from the excluded volume interaction of two polymer segments with each other; this is just proportional to the density of segments, $c(z)$, with an interaction strength per unit volume, $Tv > 0$, arising from a virial expansion, where v is the excluded volume of a segment. The energy to stretch the polymer to a height z is just $\frac{1}{2}kz^2$ multiplied by the probability, $P(z)$ to find an end at height z. In units of T, the "spring constant" $k \approx 1/(Na^2)$ where Na is the length of a completely stretched polymer chain. This form for the stretching free energy arises from the reduction in the entropy of a stretched chain; a chain that executes a random walk has an end-to-end length, R_G, which scales as $R_G \sim N^{1/2}$, while in a good solvent, $R_G \sim N^{3/5}$. On the other hand, in the brush, R_G scales linearly with N. For a more rigorous treatment of these terms see Refs. 10,11.

Brush Density

The free energy per unit area, γ, of the brush is thus written:

$$\gamma = T \int_0^h dz \left[\tfrac{1}{2}vc(z)^2 + \tfrac{1}{2}kz^2 P(z) \right] \tag{7.25}$$

A crude approximation for the end probability is to take $P(z) = c(z)/N$; *i.e.,* to assume that the ends are uniformly distributed in the region $0 < z < h$. This turns out to be reasonable near the outer part of the brush but is not a good approximation near the surface. However, since the interactions between polymer coated surfaces is mostly sensitive to the outer polymer layers, this approximation is useful. Furthermore, the stretching energy scales like z^2, which weighs most strongly the contribution of the outer part of the brush where the approximation $P(z) = c(z)/N$ is more reasonable. Minimizing the free energy with the constraint that each chain has N monomers, we consider $\partial g/\partial c(z) = 0$, where

$$g = \gamma - T\mu \int_0^h dz\, c(z) \tag{7.26}$$

The Lagrange multiplier, μ, is fixed so that $d^2 \int c(z)dz = N$, where d^2 is the area per chain on the surface. One finds

$$c(z) = \frac{1}{v} \left[\mu - \tfrac{1}{2} \left(\frac{z}{Na} \right)^2 \right] \tag{7.27}$$

where $\mu = [(3/2\sqrt{2})(v/d^2a)]^{2/3}$ and the equilibrium brush height, h^*, is determined (after minimizing with respect to c and μ) by minimizing γ with respect to h to yield

$$h^* = Na[3v/d^2a]^{1/3} \tag{7.28}$$

The polymers are strongly stretched ($h^* \sim N$) compared to their extent in solution (where their size scales with the much smaller quantity, $N^{3/5}$), although if $d^2a \gg v$ (*i.e.*, the area per chain is large), the chains are not nearly fully stretched (*i.e.*, although $h \sim Na$, h may be much smaller than Na as indicated by Eq. (7.28) in the limit that $d^2a \gg v$).

An estimate of the repulsive force between two such polymer-coated surfaces is obtained by assuming that one surface pushes against the other and changes h from its equilibrium value of h^*. For small compressions, the free energy change scales with the molecular weight N and a power of the fractional compression; for large N, this repulsive energy can overcome the attractions due to van der Waals interactions and hence stabilize the colloidal dispersion. For the simplified model described here, the free energy change per chain, Δf, as a function of $\Delta = (h^* - h)/h^*$ scales as

$$\Delta f \sim TN \left(\frac{v}{d^2a}\right)^{2/3} \Delta^2 \qquad (7.29)$$

A more accurate treatment[11] of the distribution of ends yields a similar result, except that the change in free energy is proportional to Δ^3, indicating a "softer" potential.

7.7 STRUCTURE OF COLLOIDAL AGGREGATES

Compact and Tenuous Clusters

While electrostatic and steric repulsions can sometimes prevent the van der Waals attractions between colloidal particles at short range from causing aggregation in colloidal dispersions, this is not always the case. For example, electrostatically stabilized colloids can be destabilized by the addition of salt, which screens the repulsions and promotes flocculation. Similarly, polymeric chains that are too short can be ineffective in producing steric stabilization. It is therefore of interest to consider the types of structures formed by colloidal aggregates in the limit of strong, short-range attractions between the particles. We shall consider two types of aggregates, as shown in Fig. 7.6: (i) compact aggregates, formed when the system is able to kinetically achieve equilibrium; (ii) tenuous, or fractal[2,3,13] aggregates, formed when the cluster shapes are determined by kinetics. Compact aggregates are characterized by the fact that each particle has a large number (*e.g.*, 6 for simple cubic packing, 12 for face-centered cubic packing) of nearest neighbors. They are the lowest free-energy state in a system with strong attractive interactions. Tenuous or fractal aggregates are characterized by the fact that each particle has only a small number of nearest neighbors (*e.g.*, 2) and by a local structure that is "stringy" in nature. More quantitatively, fractal structures are characterized by a scale-invariance, which implies that the number

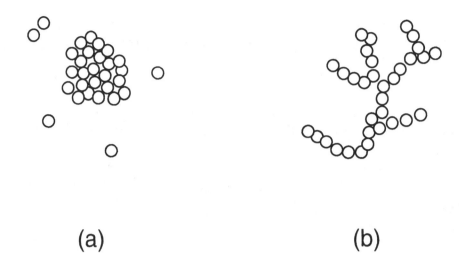

(a)   **(b)**

Figure 7.6 (a) Compact aggregate in equilibrium with a dilute colloidal dispersion. As time evolves, the aggregate grows and in equilibrium, there is macroscopic phase separation between the low- and high-density phases. (b) Schematic representation of a fractal aggregate.

of particles in a region of size R, $N(R)$, scales with a power-law relationship: $N(R) \sim R^{d_f}$, where d_f is known as the fractal dimension and $d_f < d$, where d is the dimension of space. A compact structure would, of course, result in $N(R) \sim R^d$ so that the density of particles in that region, $n \sim N/R^d$, would be a constant; it is characteristic of a fractal that the density of particles in a region of size R *decreases* as $1/R^{d-d_f}$. Had there been a characteristic length scale (other than the size of a single colloidal particle), this length scale would enter into the relationship between $N(R)$ and R and the simple, power-law scaling would no longer be valid.

Kinetics of Aggregation

Consider first a colloidal dispersion with attractive interactions where the system exists as a single phase above some critical temperature, T_c, and phase separates into two coexisting phases — one with a high density of colloidal particles and one with a low density of particles — at temperatures $T < T_c$. If one prepares the system as a uniform, single phase at $T > T_c$ and quickly lowers the temperature to $T \ll T_c$, the particles will begin to aggregate and domains of high density will form and grow in time. These aggregates will involve clusters of colloidal particles. If the time scale for two particles to stick and unstick is smaller than the time scale of the experiment, these clusters will resemble the equilibrium high-density phase and will consist of compact regions with a high density of particles. The reason these clusters are compact in equilibrium is that at temperatures $T \ll T_c$, the cluster entropy is negligible compared to the attractive interaction, which is maximized by having each particle interact with a large number of near neighbors. The system should resemble one that undergos liquid-gas phase separation and the high-density clusters are similar to the liquid "droplets" that nucleate in the gas as the system evolves toward the liquid-gas, two-phase equilibrium.

However, when the attractive interactions are much stronger than $k_B T$, the time scale for two particles to unstick, which is an exponential function of the sticking energy, may be prohibitively long. In this case, the clusters will not assume their equilibrium, compact shape since the particles that compose these aggregates are immobilized and cannot shift their positions to produce the lowest free-energy, compact cluster. The particles are effectively "stuck" whenever they meet. The resulting structure for the aggregate reflects this fact by its stringy appearance, which arises from the random sticking process and from the fact that particles are most likely to find other particles at the periphery of the cluster. This makes penetration into the cluster unlikely and it is for this reason that the clusters do not attain their equilibrium, compact shapes. Both experiments and computer simulations indicate[2,3] that these aggregates are fractal in nature. The exact value of the fractal dimension, d_f, depends on the kinetics of the aggregation. For clusters that are composed of particles that stick to each other immediately and irreversibly upon their first collision, the cluster growth rate is limited only by the diffusion of the clusters. This is known as diffusion-limited aggregation (DLA) and $d_f \approx 1.8$. Another class of aggregate is known as reaction-limited (RLA) where the particles still stick irreversibly. However, in this case, there is a kinetic barrier separating the noninteracting and strongly attractive states which is assumed to be $\gg k_B T$; the rate limiting step is thus the "reaction" — i.e., two particles may approach each other quite closely but not bind, due to this barrier. Not surprisingly, the fractal exponent is larger in this case; because the particles do not always attract when they come close, there is more opportunity for them to "fill in" the aggregate, which is more compact than that formed by DLA.

7.8 PROBLEMS

1. Force Between Charged Colloidal Particles

Explain why the force between a periodic array of charged colloidal particles is only approximately related to the charge density at the Wigner-Seitz cell boundaries — *i.e.*, show how the derivation of this relation for charged plates in Chapter 5, breaks down for an array of spheres. When might it be a particularly good or bad approximation?

2. Charge Renormalization of Rods

What is the potential and charge density as a distance of the polar radial coordinate r for a periodic array of rigid, straight rods of thickness a with a charge per unit length λ_0? To simplify the mathematics, consider a single rod in its "Wigner-Seitz" cell, which you approximate to be a cylinder of radius D, where D is related to the average separation between rods. [*Hint*: To solve the differential equation, use the transformation $v = -\ell V + \log r^2$ and $u = \log r$ and express the differential equation in terms of dv/du and a function of v, which you can then solve.] Carry out the problem analytically as far as possible; at least discuss the limiting cases. In the limit that $D \gg a$, discuss the D dependence of the effective charge on the rods as seen by charges that are at large distances from them — *i.e.*, the renormalization of the charge (the function Z^*/Z) as a function of D, and compare with the planar case discussed in the text. This model has been used to describe polyelectrolyte chains.

3. Fractal Aggregates

What is the density of an aggregate which is characterized by $N(R) \sim R^{d_f}$, where $N(R)$ is the number of particles in a region of size R? What is the real-space, density-density correlation function, $c(\vec{r})$, (the probability that there is a particle at position $\vec{r}$, given the fact that there exists a particle at $\vec{r} = 0$), of such an aggregate? What is the resulting wavevector dependence of the scattering structure factor (proportional to the Fourier transform of the density-density correlation function)?

In Ref. 14 it is suggested that fractals with a finite extent, L, can be described phenomenologically by a real-space correlation function,

$$c(\vec{r}) = \delta(\vec{r}) + A\, r^{-\alpha}\, e^{-r/L} \tag{7.30}$$

The first term comes from the self-correlation of the particle. Discuss the motivation for the second term and relate α to d_f (the fractal dimension) and d (the dimension of space). What is the structure factor for scattering from this object and discuss it in the limits of small, intermediate, and large wavevectors.

4. Polymer Brush in a Theta Solvent

A theta solvent is one where the second virial coefficient (the coefficient, v in Eq. (7.25)) — which balances the hard-core repulsion and the polymer-polymer attraction — vanishes. This occurs for many solvents only at a particular value of the temperature. (When the second virial coefficient becomes negative, the solution is unstable to phase separation and eventually, to the collapse of the individual chains into compact objects.) The free energy per unit area of a brush in a theta solvent therefore does not have any terms quadratic in the concentration and is written:

$$\gamma = T \int_0^h dz \left[\tfrac{1}{3} v_3 c(z)^3 + \tfrac{1}{2} kz^2 P(z) \right] \tag{7.31}$$

where the third virial coefficient, v_3, is positive.

Find the characteristics of the polymer brush in a theta solvent — i.e., the concentration profile and free energy. How do they differ from those in a good solvent where the second virial coefficient is positive?

7.9   REFERENCES

1. See Refs. 9,10 of Chapter 1 and R. J. Hunter, *Foundations of Colloid Science* (Oxford University Press, Oxford, 1986).

2. T. A. Witten and L. M. Sander, *Phys. Rev. Lett.* **47**, 1400 (1981).

3. D. A. Weitz, M. Y. Lin, and J. S. Huang, in *Complex and Supramolecular Fluids*, eds. S. A. Safran and N. A. Clark (Wiley, New York, 1987), p. 509.

4. C. G. de Kruif, J. W. Jansen, and A. Vrij, in *Physics of Complex and Supermolecular Fluids*, eds. S. A. Safran and N. A. Clark (Wiley, New York, 1987), p. 315.

5. J. P. Hansen and I. R. McDonald, *Theory of Simple Liquids* (Academic Press, New York, 1990), p. 95.

6. S. Alexander, P. M. Chaikin, P. Grant, G. J. Morales, P. Pincus, and D. Hone, *J. Chem. Phys.* **80**, 5776 (1984).

7. S. A. Safran, P. A. Pincus, M. E. Cates, and F. C. MacKintosh, *J. Phys. (France)* **51**, 503 (1990).

8. See the review by P. Pincus, in *Lectures on Thermodynamics and Statistical Mechanics, XVII, Winter Meeting on Statistical Physics*, eds. A. E. Gonzalez and C. Varea (World Publishing, Singapore, 1988).

9. P. G. de Gennes, *Macromolecules* **15**, 492 (1982).

10. S. Alexander, *J. Phys. (France)* **38**, 983 (1977).

11. S. T. Milner, T. Witten, and M. Cates, *Macromolecules* **21**, 2610 (1988); A. M. Skvortsov *et al.* Vysokomol. Soedin. Ser. A30, 1615 (1988).

12. P. Pincus in *Phase Transitions in Soft Condensed Matter*, eds. T. Riste and D. Sherrington (Plenum, New York, 1989).

13. B. B. Mandelbrot, *The Fractal Geometry of Nature* (Freeman, San Francisco, 1982).

14. S. K. Sinha, T. Freltoft, and J. Kjems, in *Kinetic Aggregation and Gelation*, eds. F. Family and D. P. Landau (Elsevier, Amsterdam, 1984), p. 87.

CHAPTER 8

Self-Assembling Interfaces

Micelles, Vesicles, Microemulsions

8.1 INTRODUCTION

The analysis of self-assembling dispersions such as micelles, vesicles, and microemulsions is more complex than the study of dispersions of rigid colloidal particles. Self-assembled dispersions, usually formed by the aggregation of amphiphilic molecules in solution, can vary their size and shape in response to changes in the physical properties of the system such as concentration and temperature. Thus, changes in concentration not only change the number of aggregates in solution (as in a rigid-particle colloidal system) but also their size and shape. Often, the energy scale of the interactions between self-assembled aggregates or membranes in solution is comparable to the energy scales that determine the structure of a single aggregate. This leads to a rich variety of structures and phases that allows these materials to be used in a variety of applications, such as dispersion, cleaning, and microencapsulation. Recently, there has been an explosion[1] of activity in this field as concepts and experimental approaches used to probe phase transitions in solids, liquid crystals, and polymers have been applied to self-assembled surfactant dispersions. In this chapter, we focus on the self-assembly of these systems and on the problems of the sizes and shapes of the aggregates. We first consider the case of dilute solutions of micelles, vesicles, and microemulsions where the interactions between the aggregates can be ignored. The chapter concludes with a brief discussion of random microemulsions and related spongelike phases where excluded volume interactions between the amphiphilic films are taken into account. The interplay of more complex interactions and the aggregate shape is a problem of current research[1].

8.2 MICELLES

Self-Assembly

When amphiphilic molecules are dispersed in a single solvent, such as water, the hydrophobic interactions of the hydrocarbon chains drive the molecules to self-assemble into structures where the hydrophobic tails are shielded from unfavorable interactions with the polar solvent by the hydrophilic, polar head groups. These structures typically have characteristic sizes which can be much larger than those of a single molecule (see Fig. 8.1). These structures are not covalently bonded, but rather stabilized by the weaker hydrophobic interactions; the sizes and shapes can change as a function of temperature, salinity, and/or surfactant concentration. When the spontaneous curvature of the surfactant interface (see the discussion of curvature elasticity in Chapter 6) is of the order of the amphiphile size, micellar aggregates are favored. For small spontaneous curvatures, flat bilayers or possibly vesicles are more stable than micelles. For very dilute systems, the bilayers are not necessarily stacked in a one-dimensional lamellar ordering, and a spongelike, random bicontinuous structure may be favored.

Micelles

Before considering large-scale structures such as lamellae, vesicles, and microemulsions which are well described by a bending free energy, we focus on micelles — a dispersion of surfactant in a *single* solvent where at least one dimension of the aggregate (*e.g.,* cylinder radius, disk thickness) is comparable to a molecular size. We consider a system of aggregates with N surfactant molecules per micelle (N is the aggregation number). In general, the energy of an aggregate with aggregation number N, is denoted as $E_N = TN\epsilon_N$ so that ϵ_N is the energy per surfactant in units of $k_B T$. The hydrophobic interactions favor aggregation, so we consider the case where ϵ_N is a monotonically *decreasing* function of N near $N = 1$; it may have a minimum at some larger value of $N > 1$. Treating the ensemble of aggregates in the dilute limit (*i.e.,* the total surfactant volume fraction, $\phi_s \ll 1$), we consider the free energy per (water and surfactant) molecule. (This is proportional to the free energy per unit volume if the water and surfactant molecules have equal molecular volumes.) This free energy which we denote by

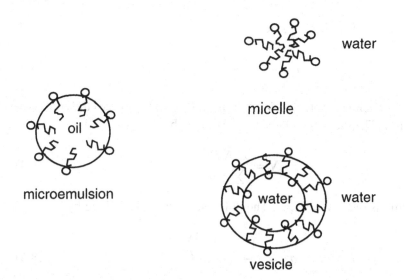

water

micelle

oil

microemulsion

water   water

vesicle

Figure 8.1 Micelles — surfactant aggregates in a single solvent; microemulsions — domains of water in oil (or oil in water) surrounded by a surfactant monolayer; and vesicles — domains of water in water (or oil in oil) surrounded by a surfactant bilayer.

f, is the sum of the energy and ideal gas entropy of each "species" (aggregates of differing N *are* distinguishable):

$$f = \sum_N \frac{P_N}{N} \left[T \left(\log \frac{P_N}{N} - 1 \right) + E_N \right] \tag{8.1}$$

Here P_N is the fraction of surfactant molecules (*i.e.*, surfactant/(surfactant + water)) that are incorporated into micelles of aggregation number N. The volume fraction of aggregates of size N is thus proportional to P_N/N. For simplicity we consider the case where the molecular volumes of the surfactant and the solvent are equal, so that $\sum_N P_N$ which is the fraction of surfactant molecules in the system, can also be written as

$$\sum_N P_N = \phi_s \tag{8.2}$$

Otherwise, we can just consider ϕ_s to be the number fraction (or mole fraction of surfactant molecules). Minimizing the free energy with respect to P_N subject to the constraint of Eq. (8.2), results in

$$\epsilon_N + \frac{1}{N} \log \frac{P_N}{N} = \mu \tag{8.3}$$

where μ is the chemical potential *in units of* $k_B T$, which enforces the conservation constraint. The mean micelle size, $\bar{N}$ is given by

$$\bar{N} = \frac{\sum N P_N}{\sum P_N} \tag{8.4}$$

where the sum is over $N = 1...\infty$. The polydispersity, σ, is defined by

$$\sigma^2 = \frac{\left\langle \left(N - \bar{N}\right)^2 \right\rangle}{\bar{N}^2} \tag{8.5}$$

where

$$\left\langle N^2 \right\rangle = \frac{\sum N^2 P_N}{\sum P_N} \tag{8.6}$$

Critical Micelle/Aggregate Concentration

Noting that both ϵ_N and the chemical potential are dimensionless, Eq. (8.3) indicates that the fraction of amphiphile in a micelle of aggregation number N can be written as

$$P_N = N e^{N(\mu - \epsilon_N)} \tag{8.7}$$

When $\mu < min\{\epsilon_N\}$ (which corresponds to small values of ϕ_s), P_N is exponentially small for large N. The most probable aggregates are monomers of size $N = 1$. As μ is increased, however, the number of these monomers cannot increase indefinitely, because of the conservation constraint. When the difference $\mu - \epsilon_N$ becomes small, the larger aggregates become more probable. This is because the entropy of mixing, which favors the small aggregates becomes less important. The **critical micelle concentration (CMC)** (or critical volume fraction, ϕ_c) is defined (somewhat arbitrarily) by the condition

$$\phi_c - P_1 = P_1 \tag{8.8}$$

i.e., the fraction of amphiphile in micelles of $N > 1$ is equal to the fraction of amphiphile in monomers. For values of ϕ_s greater than ϕ_c the number of monomers remains approximately constant, while the number of aggregates increases. The details of this process depends on the dependence of the energy ϵ_N on the aggregation number.

Similar considerations apply to the self-assembly of other aggregates such as microemulsions, lamellae, and vesicles. At small volume fractions of surfactant, the probability to find these aggregates is small; most of the surfactant is incorporated in smaller objects (such as isolated surfactant molecules or small micelles). Above a critical aggregation concentration (CAC), the probability to find the large aggregates is of order unity; as the amount of surfactant is increased, the volume fraction of small objects remains constant, while the number of large aggregates (*e.g.*, number of lamellae, vesicles, microemulsion domains) increases. The approximation discussed in Chapter 6, where only the local interfacial energy of the large-scale aggregate is considered, is appropriate in this higher concentration regime, where the small objects can be neglected.

Spherical Micelles

A simple example is one where the hydrophobic interactions result in $\epsilon_N < \epsilon_1$ for all $N > 1$, but the packing constraints on the chains and heads result in a minimum energy for a finite value of $N = M$ (*i.e.*, $\epsilon_M < \epsilon_N$ for $N \neq M$). (Below, a curvature energy model is discussed; this model can also be used to motivate, but not to calculate in detail, a study of micellar sizes and shapes.) If this minimum is deep (*i.e.*, ϵ_N rises sharply compared to $k_B T$ around $N = M$), the distribution of micelles will be nearly monodisperse. In this approximation, one can consider monomers and micelles of aggregation number M only. At small values of ϕ_s (or equivalently, at small values of μ), $P_1 \gg P_M$ ($M > 1$); almost all the surfactant exists as monomers and the number of micelles is exponentially small. The requirement that all amphiphiles have the same chemical potential in equilibrium, Eq. (8.2) and the definition of the CMC (where $P_1 = P_{1_c}$, $P_M = P_{M_c}$, $\mu = \mu_c$, $\phi = \phi_c$) allows us to calculate the populations of micelles and monomers at the CMC and we find

$$P_{1_c} = e^{\mu_c - \epsilon_1} = \tfrac{1}{2}\phi_c \tag{8.9}$$

Similarly,

$$P_{M_c} = M\, e^{M(\mu_c - \epsilon_M)} = \tfrac{1}{2}\phi_c \tag{8.10}$$

Eliminating the chemical potential from these equations gives an expression for the CMC:

$$\phi_c = 2\,(M)^{1/(1-M)} \left[e^{\epsilon_1 - \epsilon_M} \right]^{M/(1-M)} \sim e^{\epsilon_M - \epsilon_1} \tag{8.11}$$

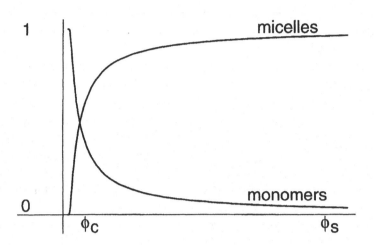

Figure 8.2 Fraction of surfactant in monomers (P_1/ϕ_s) and in micelles (P_M/ϕ_s) for a model where there is only one preferred size $\epsilon_M \ll \epsilon_N$ for $N \neq M$. The critical micelle concentration is denoted by ϕ_c and ϕ_s is the surfactant volume fraction.

Thus, if the preferred packing favors a value of $M \gg 1$, the CMC is essentially determined by the difference between the energy per amphiphile of the monomers and of the aggregates. Systems with strong hydrophobicity show very small values of the CMC, typically at volume fractions of $10^{-8} - 10^{-4}$. For values of $\phi_s > \phi_c$ (or equivalently, values of $\mu > \mu_c$), the number of micelles increases as ϕ_s increases, while the number of monomers stays approximately constant, as indicated in Fig. 8.2.

Cylindrical Micelles

While molecules that prefer spherical aggregates are constrained by geometrical packing to assemble into micelles with a fixed value of $N \approx M$ (the packing constraints force ϵ_N to have a deep minimum for $N \approx M$), amphiphiles which prefer cylindrical aggregates show a broad distribution of sizes. This is because the energy depends only on end-effects such as the cylindrical end-caps; the energy per amphiphile in the cylinder interior is independent of N. A simple model for cylindrical micelles considers an energy per amphiphile of the form:

$$\epsilon_N = \epsilon_\infty + \alpha/N \tag{8.12}$$

where α is the end energy per surfactant in units of $k_B T$. We consider the case where it is *energetically* favorable to form cylinders so $\alpha > 0$ represents the additional energy of the material in the end-caps. Assuming that this form holds for all N, we eliminate the chemical potential in favor of the quantity P_1 (see Eq. (8.7)) and find

$$\phi_s = \sum_N P_N = \sum_N N (P_1 e^{\alpha})^N e^{-\alpha} \tag{8.13}$$

Using the relation:

$$\sum_N N x^N = \frac{x}{(1-x)^2} \tag{8.14}$$

we find that

$$P_1 = \frac{(1 + 2\phi_s e^{\alpha}) - \sqrt{1 + 4\phi_s e^{\alpha}}}{2\phi_s e^{2\alpha}} \tag{8.15}$$

At low volume fractions, $\phi_s e^{\alpha} \ll 1$, almost all the surfactant is in monomers, $P_1 \approx \phi_s$. Thus, the CMC is defined by $\phi_c \approx e^{-\alpha}$. For concentrations much larger than ϕ_c, we find that $P_1 \approx e^{-\alpha} \approx \phi_c$ and the distribution of aggregates is determined by

$$P_N = N \left(1 - \phi_s^{-1/2} e^{-\alpha/2} \right)^N e^{-\alpha} \tag{8.16}$$

The most probable value of the aggregation number (where $\partial P_N / \partial N = 0$) is defined as

$$M = \sqrt{\phi_s e^{\alpha}} \tag{8.17}$$

For large N, the distribution is approximated by

$$P_N \sim N e^{-N/M} \tag{8.18}$$

The distribution is polydisperse with a large variety of aggregates with $N \le M \gg 1$ in contrast to the spherical case where the probability falls to zero for $N = M \pm m$ where m is typically a number of order unity. This polydispersity, as well as the dependence of the most probable size on the square root of the concentration is linked to the one-dimensional nature of the aggregate, Eq. (8.12). If the finite size corrections go to zero more slowly than $1/N$ (e.g., $1/\sqrt{N}$ for disks), there are very few aggregates where N is large; the system partitions between monomers and infinite flat bilayers, instead of making a polydisperse distribution of large disks. Of course, if there are specific packing constraints (see Ref. 2) that prefer a disk of a particular size, then the distribution will be peaked about this size, similar to the spherical case.

8.3 VESICLES

Equilibrium Vesicles

Unilamellar vesicles consist of a surfactant bilayer that separates an inner region of a fluid (usually water) from a continuous phase of the same fluid. Industrial and biological applications such as cleaning, catalysis, and microencapsulation for drug delivery depend on a simple and controlled method for the generation of vesicles with a well-defined average size. In addition, vesicles are often studied as models for biological membranes. Although vesicles often form spontaneously in vivo, they rarely form as the equilibrium structure of simple surfactant-water systems. Nonequilibrium methods, such as sonication of lamellar, liquid crystalline phases, are usually necessary to obtain a metastable phase of vesicles, which may equilibrate back into the multilamellar, liquid crystalline structure. Recently, however, there have been reports[3,4,5] of a general method for producing equilibrium phases of vesicles of a controlled size. The vesicles form spontaneously upon mixing simple surfactants with either oppositely charged head groups and/or different chain lengths. Here, we shall examine the stability of vesicles composed of a *single* surfactant. We shall see that they are *entropicly* stabilized, the lowest curvature energy state (see Chapter 6) being a flat, lamellar bilayer. The unusual stability of vesicles formed by mixed surfactants is discussed theoretically in Ref. 6.

Curvature Energy and Frustration in Vesicles

We consider the curvature energy of spherical vesicles. As a first approximation, we assume that the area per molecule is the same in each of the two monolayers that compose the vesicle bilayer; we therefore just add the bending energies of each monolayer. The total bending energy per unit area of the midplane between the two monolayers that compose the bilayer can be obtained from Eq. (6.15). Noting that for spheres, the two curvatures $\kappa_1 = \kappa_2 = c$, we write:

$$f_c = 2k \left[(c + c_0)^2 + (c' + c'_0)^2 \right] + \bar{k} \left[c^2 + c'^2 \right], \qquad (8.19)$$

where c_0 and c'_0 are the spontaneous curvatures of the inner and outer monolayers respectively and c and c' are the actual curvatures of the inner and outer monolayers. We shall use the convention that the inner monolayer of the vesicle has positive curvature and the outer monolayer, negative curvature. To a good approximation (see the discussion of parallel surfaces in Chapter 1), $c \approx -c'$, where terms of order $c\delta$ have been neglected (δ is the monolayer thickness). For the case of single surfactant systems, in the limit of small curvatures, the chemical properties of the two layers are the same and hence the spontaneous curvatures, of the inner and outer layers are equal ($c_0 = c'_0$). In this case, the terms in Eq. (8.19) that are linear in curvature vanish and the minimum of f_c with respect to c implies that $c = 0$ — *i.e.,* flat bilayers are the lowest bending energy state. The physical origin of the instability of the vesicle to the flat bilayer is the frustration of the curvature energy in the spherical state; if the curvature is chosen so that, for example, the outer layer is at its spontaneous curvature, the inner layer is frustrated and vice versa. Thus the only way for the system to minimize its curvature energy is by forming a flat bilayer.

Of course, the two layers of the bilayer do not have areas per molecule which are exactly equal, nor are their curvatures equal and opposite; the areas are self-adjusting in equilibrium[7]. This can result in a vesicle that is more stable than a flat, lamellar bilayer since it may be that the favorably curved layer (*e.g.,* the outer layer) has a smaller value of the area per molecule and hence more molecules in it than the inner layer. However, such corrections are only important for systems with large spontaneous curvatures[7] and the resulting vesicles would have sizes comparable to molecular lengths. It therefore may turn out that micelles may be more stable than either vesicles or lamellae. In what follows, we consider the spontaneous formation of large vesicles whose sizes are much greater than a surfactant size and hence neglect these effects.

Stability Bound on Entropy Stabilized Vesicles

We have seen that for *large* vesicles composed of a single amphiphile, the lamellar phase has lower curvature energy than the vesicle. However, since the vesicles have a finite size, they have an entropy of mixing that is much larger than that of the lamellar phase. As a bound on the stability of the vesicle phase, we consider the dilute limit where the excluded volume interactions of the vesicles can be neglected. We thus include the translational entropy and calculate the distribution of vesicle sizes obtained by minimizing the total free energy per unit volume, f, which is

$$f = T \sum_N \left(n_N \left[\log(n_N v) - 1 \right] + n_N k_t + n_N N \mu \right) \qquad (8.20a)$$

Here, v is a cutoff volume, n_N is the number of vesicles per unit volume of aggregation number N ($N \sim c^{-2}$), $k_t = 8\pi(2k+\bar{k})/T$ is the energy per vesicle in units of $k_B T$, and μ is a Lagrange multiplier which accounts for the conservation of surfactant and is determined from

$$\sum_{N} v n_N N = \phi_s. \tag{8.20b}$$

where ϕ_s is the surfactant volume fraction.

Vesicle Distribution

Minimizing the free energy of Eq. (8.20a) with respect to n_N yields the vesicle distribution,

$$n_N = \frac{1}{v} e^{-k_t - \mu N}. \tag{8.21}$$

The Lagrange multiplier, μ is evaluated from the conservation constraint and is proportional to $\left[\phi_s e^{k_t}\right]^{-1/2}$. This distribution is *not* Gaussian about some average size (usually determined by the energetics), since the minimum energy state is a vesicle with curvature $c = 0$ corresponding to the infinite aggregation number of a flat bilayer. The average vesicle size, $\bar{R} = \sum N^{1/2} n_N / \sum n_N$ is proportional to $1/\mu^{1/2}$ and diverges exponentially as the bending stiffness increases. Of course, at finite surfactant volume fraction, the average size is finite. However, the value of ϕ_s at which these vesicles can still remain dilute is exponentially small. To see this, we determine the volume fraction enclosed by the vesicles, Φ, which scales as

$$\Phi \sim \sum_{N} N^{3/2} n_N \sim \phi_s \bar{R}. \tag{8.22}$$

Since $\bar{R} \sim \left[\phi_s e^{k_t}\right]^{1/4}$, we see that the vesicles are overpacked (*i.e.,* $\Phi > 1$), unless ϕ_s is exponentially small. For stiff membranes, where $k_t \gg 1$, this implies that the vesicles can only exist as a dilute solution for surfactant volume fractions, $\phi_s \ll e^{-k_t}$; for larger volume fractions, the lowest free energy state is probably lamellar. Including the effects of the renormalization of the bending modulus to lower values due to long-wavelength fluctuations (see Chapter 6), changes the distribution of Eq. (8.21) from a simple exponential to the product of a power law in N and an exponential[8].

We therefore conclude that for single surfactants the lamellar phase is usually more stable than the spherical vesicle phase. Exceptions to this behavior can occur for (i) the case of small vesicles, comparable to the surfactant size, where the detailed energetics may stabilize the system — although the micellar phase may ultimately be even more stable, or for (ii) the case of extremely small surfactant volume fractions, or small values of the bending energy compared with $k_B T$, where the entropy of mixing can stabilize a polydisperse distribution

of large vesicles. In contrast with the trend for single surfactants, mixed surfactants can form a phase of large vesicles with a well-defined average size that
is determined[6] by the curvature energy of the interacting system. In the mixed
system, interactions between the two species can result in a spontaneous curvature of a mixed layer that is very different from the spontaneous curvatures
of films of either species. If the mixing is not the same in the two monolayers
that comprise the bilayer, the spontaneous curvatures of these two monolayers
may be different enough to relieve the curvature frustration that exists in vesicles
composed of a single surfactant species.

8.4 MICROEMULSIONS

Microemulsions vs. Solutions

Mixtures of amphiphiles, water, and oil can form[9,10,11] structures that have
only microscopic correlations between the components (*e.g.*, a three-component
solution) as well as long-range ordered structures (*e.g.*, lyotropic liquid crystals).
The term **microemulsion**, in its most general use, connotes a thermodynamically
stable, fluid, oil-water-surfactant mixture. In practice, microemulsions are taken
to consist of structures with intermediate-range correlations. The oil and water
regions are fairly well separated by the organized surfactant monolayers[12]. There
can be long-range correlations between the oil and water molecules in that they
are separated on length scales of the order of hundreds of Angstroms. In addition,
there are long-range correlations among the surfactant molecules, which self-
assemble into a monolayer film at the set of internal water-oil interfaces. In
this respect, microemulsions are differentiated from three-component solutions.
However, the set of interfaces that comprise the microemulsion do not show
long-range order comparable to that found in lyotropic liquid crystals.

It is precisely for these reasons that the understanding of microemulsions is
both interesting and difficult: interesting, because the structure can be idealized
as a set of interfaces, and difficult, because there is no long-range order of these
interfaces. Although many workers[1,13,14] have implicitly used a "phenomenological" approach to analyze globular (mostly spherical) microemulsions, the
application of the physics of random surfaces to microemulsions has only recently been developed[15,16,17]. The thermal fluctuations of the surfactant film
have been considered[18,19] in modeling both the thermodynamics and spatial correlations of bicontinuous microemulsions[20]. One regards the oil and water as
continuum liquids; the interfacial surfactant layer is treated as a flexible sheet.
For systems where the surfactants pack in a condensed liquid state, the dominant
energy is the bending or curvature energy of the monolayer. While this energy
is minimized by a droplet or domain with a given curvature (determined by the

molecular details of the surfactant packing at the interface), the entropy tends to randomize the structure.

Microscopic Lattice Models

An alternative approach is based on the construction of microscopic lattice models in which a cell contains only a small number of molecules[14,21]. This point of view, which focuses on the microscopic interactions between the water, oil, and surfactant molecules, is well suited to describe three-component solutions and their relation to microemulsions. The microscopic approach, though of important fundamental interest, may be more difficult to implement than the phenomenological one. In particular, a microscopic model of microemulsions must produce structural organization on a length scale much larger than that of the molecular (or lattice) size. In contrast, the interfacial point of view presented here is tailored to focus on the microstructure and its relation to the phase behavior of microemulsions. This is because the length scales of interest for the problems of microstructure and scattering can be in the range of several tens to several hundreds of Angstroms for some microemulsions, where the surfactants are very insoluble in either the water or the oil. These length scales are best treated by the continuum theory presented here; microscopic theories would have to include correlations between thousands of molecules to accurately model the interfaces at these length scales. In contrast, the interfacial models presented here assume that the strong correlations needed to self-assemble the surfactant molecules at the internal water-oil interfaces are always present.

Competition Between Different Structures

We begin by considering the bending energy per unit area for a curved surfactant monolayer at a water-oil interface with principal curvatures (see Chapter 1) κ_1 and κ_2.

$$f_c = \tfrac{1}{2}k(2H - 2c_0)^2 + \bar{k}K, \tag{8.23}$$

where the average curvature is $H = \tfrac{1}{2}(\kappa_1 + \kappa_2)$ and the Gaussian curvature $K = \kappa_1\kappa_2$; c_0 is the spontaneous curvature, k is the bending modulus, and $\bar{k}$ is the saddle-splay modulus. We shall find the minimum bending energy states, which will determine the shapes of microemulsion domains. This is relevant for systems where the bending energy $k \gg k_B T$ so that fluctuations of the interface can be neglected.

The shape that minimizes the bending energy for negative values of $\bar{k}$ is one of monodisperse spheres of radius $\tilde{c}_0^{-1}$, where

$$\tilde{c}_0 = c_0 \left[1 + (\bar{k}/2k)\right]^{-1} \tag{8.24}$$

However, such a configuration is only possible for particular values of the water, oil, and surfactant concentrations because of the incompressibility assumption described earlier. These constraints require that the total surface area of all of the interfaces in the system is fixed (in a single-phase system) by the surfactant concentration. Similarly, the total volume enclosed by the polar-head side of the surfactant film is fixed by the total concentration of water. Simple geometry indicates that the ideal curvature of $c = \tilde{c}_0$ for spheres can only be obtained when the surfactant/water ratio (for water internal globules) satisfies $\phi_s/\phi_w = 3\delta\tilde{c}_0$, where δ is a molecular size. Any other ratio of the concentrations can result in a system of monodisperse spheres, but with a curvature that is not equal to $\tilde{c}_0$. Therefore, even in the absence of thermal fluctuations, a large variety of structures is possible because of the competition between the tendency to minimize the bending energy (which prefers spheres of curvature $\tilde{c}_0$) and the necessity to satisfy the incompressibility constraints.

For systems where f_c dominates the free energy, the shape transitions due to the competition between the bending energy and the incompressibility can be simply calculated. Assuming a monodisperse set of globules, the incompressibility conditions are given by

$$nA\delta = \phi_s \tag{8.25a}$$

and

$$nV = \phi_w \tag{8.25b}$$

where n is the density of globules, δ is the surfactant size, and A and V are the surface area and volume of the globules respectively. The bending energy for globules of different shapes can be calculated and compared to determine the microemulsion shape. Using the constraints given by Eq. (8.25) in the expression for the bending energy per unit area, f_c, one finds the following expressions for the relative bending energies per unit area of spheres (Δf^s), and infinite cylindrical (Δf^c) and lamellar structures (Δf^l):

$$\Delta f^s = 2\tilde{k}\,\tilde{c}_0^2\left[(1 - 1/r)^2 - 1\right] \tag{8.26a}$$

$$\Delta f^c = 2\tilde{k}\,\tilde{c}_0^2\left[\frac{9}{16r^2}(1 + \bar{k}/2k)^{-1} - \frac{3}{2r}\right] \tag{8.26b}$$

$$\Delta f^l = 0 \tag{8.26c}$$

In Eq. (8.26), a constant that is independent of the sphere size — or the concentrations — has been subtracted from each of the energies so that the bending energy of the lamellae is zero. The volume-to-surface ratio of the structure is proportional to $\rho = 3\delta\phi_w/\phi_s$, determined by the constraints. The ratio of the two characteristic lengths is denoted by $r = \tilde{c}_0\rho$; these lengths are the volume-to-surface ratio, ρ (fixed by the concentrations), and the spontaneous radius of curvature, $\tilde{\rho}_0 = \tilde{c}_0^{-1}$ (fixed by the form of the bending energy). The modulus $\tilde{k}$ is defined by

$$\tilde{k} = k + \bar{k}/2 \tag{8.27}$$

For $r > 1$, the energy is minimized by a phase of spherical globules with $\rho = \tilde{\rho}_0$, coexisting with excess internal phase (**emulsification failure**) or with a dilute, lamellar phase. This is because the lowest energy state of the system is one of spheres of radius $\rho = \tilde{\rho}_0$; if excess water is added, it is simply rejected into another phase. The effects of the entropy of mixing on this instability can be easily incorporated by adding a term to the free energy which represents the entropy of a dilute "gas" of spherical droplets and by finding the value of ρ at which the free energy is a minimum[14]. The result yields an expression for $\rho/\tilde{\rho}_0$ along the emulsification-failure phase boundary which depends on the volume fraction of droplets:

$$\rho/\tilde{\rho}_0 = 1 + \frac{T}{8\pi\tilde{k}} \log(v_0 n) \tag{8.28}$$

where v_0 is a cutoff volume, and n is the number density of drops. As r is decreased, by a reduction of either the ratio of ϕ_w/ϕ_s (for water internal systems) or of c_0, first-order phase boundaries separate the regimes of different shapes. The regions where the different shapes (spheres, cylinders, lamellae) have the lowest bending energy are shown in Fig. 8.3 as a function of the ratio $x = -\bar{k}/2\tilde{k}$ and $r = \rho/\tilde{\rho}_0$. Note again that this simplified stability diagram is indicative of the structures in the dilute limit where the interactions between microemulsion droplets can be neglected. For systems where these interactions are important, one can treat the system as a colloidal dispersion so long as there are no shape changes as parameters such as concentration or temperature are varied.

The calculation of the shape stability whose results are shown in Fig. 8.3, indicates that these regions include domains of two-phase coexistence of spheres and cylinders, for example. The details of this coexistence may also depend on the entropic contributions to the free energy. However, it is important to note that a comparison of the bending energies of spheres and cylinders implies that for $\bar{k} = 0$ the two shapes are degenerate, albeit at different values of r. This implies that spheres and cylinders coexist in a two-phase equilibrium with no single phase of spheres when $\bar{k} = 0$ (see Fig. 8.3). Thus, a single phase of

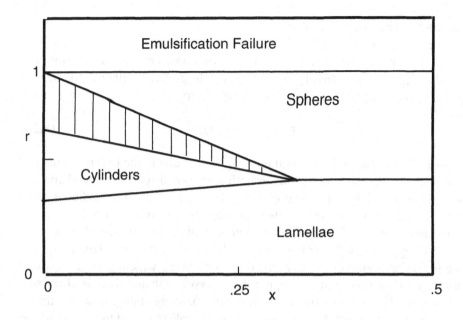

Figure 8.3 Phase stability of spherical, cylindrical, and lamellar phases. For simplicity, we show only the crossings with the lamellar free energy; the cross-hatched region is a two-phase coexistence of spheres plus cylinders. The parameter $r = 3\tilde{c}_0\delta\phi_w/\phi_s$, where δ is the surfactant size, $\tilde{c}_0$ the spontaneous curvature defined in the text, and ϕ_s (ϕ_w) the surfactant (water) volume fractions for water in oil systems; $x = -(\bar{k}/2\tilde{k})$.

spheres is only stable when $\bar{k}$ is negative and nonzero; the size of this one-phase region is related to magnitude of $\bar{k}$. The cylindrical phase is stable in a region of r and $\bar{k}$ where it best accommodates the volume and surface constraints and still maintains an average radius of curvature close to $\tilde{\rho}_0$; for small saddle-splay elasticity, $\bar{k}$, there is little cost in generating an anisotropic structure.

For large values of the saddle-splay modulus ($x > 1/3$), only shapes with identical orthogonal radii of curvature (spheres and lamellae) are present. Finally, at small values of $\tilde{c}_0$, for all values of x, the stable shape is lamellar; there is no preferred globule since there is no energetic tendency to bend toward either the oil or the water. One should note that since entropy effects generally stabilize smaller objects, inclusion of the entropy of mixing would enlarge the region of stability of the spherical droplet phase compared with the cylindrical and lamellar structures.

Microemulsion Fluctuations

For systems where the bending energy is comparable to $k_B T$, thermal fluctuations can modify the shapes of the equilibrium microemulsion domains. The probability, P, that an arbitrary deformation of the globules will occur in thermal equilibrium is proportional to the Boltzmann factor

$$P \sim \exp(-F_b/T) \tag{8.29}$$

where F_b is the change in the total bending energy due to the deformation. The magnitude of thermal fluctuations in both size (polydispersity) and shape can then be calculated. For spherical globules, the main effect of these fluctuations on the ensemble of nominally spherical droplets is to induce a polydispersity in size and shape[22,23,24]. However, the integrity of the description of the system as one of approximately monodisperse spheres is maintained. This is not the case for the cylindrical[25,26] and lamellar[16] structures, which have at least one very long length scale that can support long-wavelength undulations of the surfactant film. These undulations can have little bending energy cost and thus a large Boltzmann weight. For systems that are flexible (k/T not too large) and/or dilute enough, the zeroth order picture of infinite, rigid cylinders, or liquid crystalline, lamellar order may be radically modified by these fluctuations. In the case of cylindrical structures, this can result in a disordered phase characterized by wormlike, flexible tubes that can have properties very similar to those of polymers in solution[25,26]. In the case of the lamellar structures, entropic fluctuations can lead to a "melting" of the sheets into a disordered, spongelike phase.

8.5 SPONGELIKE AND BICONTINUOUS PHASES

Spongelike phases in surfactant-water or surfactant-oil systems (the so-called L3 phases where the surfactant forms a bilayer separating the "inside" and "outside" of the sponge) or **bicontinuous phases** in microemulsions (where the surfactant monolayer separates oil and water domains) often appear in the phase diagrams in close proximity to lamellar phases. It is therefore reasonable to assume that in such a microemulsion the spontaneous curvature, c_0, of the monolayer is close to zero since at larger values of c_0, spherical or cylindrical structures might be expected. For a bilayer composed of a single surfactant, symmetry dictates that the spontaneous curvature of the bilayer is zero (see the earlier discussion of vesicles). Thus, the understanding of these phases has focused on the properties of amphiphilic interfaces with zero spontaneous curvature. Since the structures of the single solvent, L3 phases, and the bicontinuous microemulsions (when $c_0 = 0$) are quite similar, we shall focus on the specific case of the

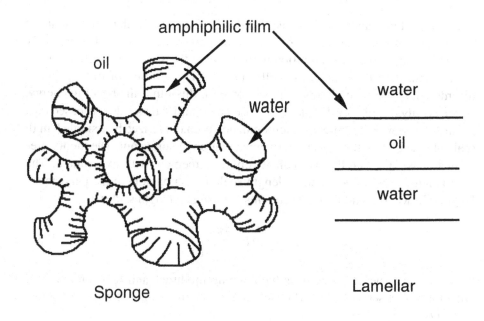

Figure 8.4 Schematic representation of spongelike bicontinuous structure in water-oil microemulsions compared with an ordered lamellar structure. A similar picture applies to an amphiphilic bilayer separating "inside" and "outside" water or oil domains for systems with a single solvent (L3 phases).

microemulsions where the inside and outside of the sponge are composed of water and oil domains. A schematic representation is shown in Fig. 8.4.

Although the understanding of the structures and phases in these systems is still a matter of current research, three major theoretical approaches have emerged. The first approach[14,18,19,27], which focuses on the proximity of the spongelike phases to lamellar phases in the phase diagram, emphasizes the role of the entropy of mixing of the water and oil regions in the bicontinuous structure. This lowers the free energy of the bicontinuous structure compared with that of the ordered, lamellar phase and stabilizes the spongelike structure. Conservation of surfactant in a lamellar phase, implies that the spacing, d, between the nearly flat surfactant sheets increases as the surfactant volume fraction, ϕ_s, is decreased: $d \sim 1/\phi_s$. When this spacing increases to the size of the natural persistence length of the monolayer (see Chapter 6), the lamellar structure "melts" into the disordered, bicontinuous, spongelike phase. While the curvature energy of the sponge is higher than that of the nearly flat lamellae, the renormalization of the bending energy by thermal fluctuations (see Chapter 6) results in a softer effective bending modulus at small surfactant volume fractions; this renormalized curvature modulus is of order $k_B T$ just when the length scale of the sponge (or lamellar phase) is close to the persistence length. It is precisely in this low surfactant concentration regime where the spongelike structures are more stable than the ordered, lamellar phases. The entropy of mixing in the disordered, spongelike phases can overcome, in this limit, the bending energy, which is only of order $k_B T$. The theory used to quantify these ideas is based on a mean-field approach where the microemulsion is characterized by a single length scale obtained from the constraint of conservation of surfactant. The theory predicts a phase diagram that is qualitatively in agreement with experiment. The microemulsion is stable when its length scale is comparable to the persistence length of a weakly undulating sheet; this persistence length scales as

$$\xi_k = a \exp\left[\frac{4\pi k}{\alpha T}\right] \tag{8.30}$$

where a is a molecular size, k is the bending modulus, and α is a number of order unity. Conservation of surfactant implies that in the microemulsion regime, $\phi_s \sim 1/\xi_k$.

A related approach[28] uses the properties of Gaussian random surfaces[29], considers both the bending energy and entropy and generalizes the previous theory to allow for a continuum of length scales. The microemulsion is characterized by an average length scale, but the fluctuations about this average can be significant and even lead to domains that are locally anisotropic, as shown in Fig. 8.5. The bending energy controls the distribution of lengths in this model. It is important to note[28] that in this model, it is not the persistence length of the lamellae that controls the structure, but a length that scales as k/ϕ_s, where k is the bending constant. Although this picture very successfully describes the scattering structure factor and presents a reasonable picture of the real-space structure, its effectiveness at predicting the phase diagram and interfacial tensions is a matter of current research.

The third approach to understanding these spongelike phases focuses[30,31] not on the entropy of the disordered structure, but on the stabilization of the sponge by the saddle-splay curvature modulus, $\bar{k}$ (see Eq. (8.23)). As shown in several examples in Chapter 6, $\bar{k}$ is usually negative (thereby favoring isotropic curvatures such as those characterizing spheres or lamellae); its value can be tuned by variations in the chemistry of the surfactant film. When it is significantly positive, saddle-shaped regions, where the two principal curvatures have opposite signs[32], are preferred and the spongelike structure is stabilized. Of course at higher surfactant volume fractions or low temperatures, one would expect this approach to predict an ordered, periodic, saddle-shaped structure[32]. In this picture, it is the saddle-splay modulus $\bar{k}$, rather than the bending modulus, k that controls the structures and phases. A more quantitative version of these

Figure 8.5 Two-dimensional cut through a bicontinuous microemulsion comprising water (white) and oil (black) separated by a surfactant monolayer. The same picture can also represent the "inside" and "outside" of the so-called L3 sponge phase where a surfactant bilayer separates inner and outer regions of a single solvent. Note that the domains have a well-defined length scale. The details of the model used to generate this representation are discussed in Ref. 28.

ideas which incorporates the bending energies and fluctuations will be useful in comparison with experiment.

Example: Random Surfaces

Consider a surface that is defined (in the implicit representation) by $\psi(\vec{r}) = \alpha$; *i.e.*, by a certain contour (level-cut) of a field, $\psi(\vec{r})$. One can[28,29] associate negative values of ψ with the water domains of a microemulsion (or with the inside of the spongelike L3 phases of a symmetric bilayer) and positive values of ψ with the oil domains of a microemulsion (or with the outside of the structure in the L3 phase of a symmetric bilayer). A microemulsion with equal volumes of water and oil would thus have $\alpha = 0$.

Calculate the thermodynamic average of the surface-averaged mean curvature, H, the mean-square curvature, H^2, and the Gaussian curvature, K, of the surface if the Fourier transform of the "random" field, $\psi(\vec{q})$ is described by a Gaussian probability distribution, $P\left[\{\psi(\vec{q})\}\right]$. Perform the calculation for the case where the different wavevectors, $\vec{q}$, are uncoupled so that one can write the probability distribution for a single mode as

$$P\left[\psi(\vec{q})\right] \sim \exp\left[-G(\vec{q})|\psi(\vec{q})|^2\right] \tag{8.31}$$

where the proportionality constant is given by the normalization. Discuss how this can be used to develop a thermodynamically and structurally consistent theory of the aforementioned phases.

The local mean curvature and Gaussian curvature in the implicit representation of a surface are given in Eqs. (1.105,1.107), where in our problem, $F(x, y, z)$ corresponds to $\psi(\vec{r})$. To calculate the statistical averages, one must find the probability distribution function for the field, ψ and its first and second derivatives at any given point since the curvatures are *local* functions of these quantities. For a system with no long-range order, these averages are independent of the position in space, $\vec{r}$, and it suffices to calculate them at $\vec{r} = 0$. The probability distribution function for the field and its derivatives is obtained from Eq. (8.31) by using the relation

$$P\left[A\left[\psi(\vec{r})\right] = A_0\right] = \prod_{\vec{q}} \int d\psi(\vec{q})\,\delta\left(A\left[\psi(\vec{r})\right] - A_0\right) P\left[\psi(\vec{q})\right] \tag{8.32}$$

Here, $A\left[\psi(\vec{r})\right]$ is any function of ψ or its derivatives and Eq. (8.32) is the probability that this function has a given value, A_0. In our case, we need the joint distribution of the field and its first two derivatives ($\psi_i = \partial\psi/\partial r_i$, $\psi_{ij} = \partial^2\psi/\partial r_i\partial r_j$ where $\vec{r} = (x, y, z)$), evaluated at the surface that is defined by $\psi = \alpha$:

$$P\left[\alpha, \vec{v}, \beta_{ij}\right] = P\left[\psi = \alpha, \nabla\psi = \vec{v}, \psi_{ij} = \beta_{ij}\right] \tag{8.33}$$

where the vector $\vec{v}$ and the matrix β_{ij} are the values of the derivatives.

This probability distribution is evaluated by using Eq. (8.32) and the relationship between the Fourier modes and the real-space functions enters in the argument of the delta function in Eq. (8.32). For example, one writes

$$\delta\left(\psi(\vec{r})\right) = \delta\left(\frac{1}{\sqrt{L^3}} \sum_{\vec{q}} \psi(\vec{q}) \, e^{i\vec{q}\cdot\vec{r}}\right) \tag{8.34}$$

where L is the system size; similar expressions are used for the derivative terms. The integral in Eq. (8.32) is performed by using the exponential form of the delta function

$$\delta(x) = \frac{1}{2\pi} \int_{-\infty}^{\infty} d\omega \; e^{i\omega x} \tag{8.35}$$

Since the argument of the delta function is always linear in $\psi(\vec{q})$, the integrals with the probability distribution involving $G(\vec{q})$ are Gaussian and can be performed. This is most easily done at $\vec{r} = 0$. The result can be written compactly in matrix notation as

$$P\left[\alpha, \vec{v}, \beta_{ij}\right] = P_0 \exp\left[-\frac{v^2}{2\sigma_v^2} - \sum_{i<j} \frac{\beta_{ij}^2}{2\sigma_{ij}^2} - U\right] \tag{8.36}$$

where P_0 is a normalization. The covariance matrix elements are given by

$$\sigma_v^2 = \frac{1}{3L^3} \sum_{\vec{q}} \frac{q^2}{G(\vec{q})} \tag{8.37}$$

$$\sigma_{ij}^2 = \frac{1}{L^3} \sum_{\vec{q}} \frac{q_i^2 \, q_j^2}{G(\vec{q})} \tag{8.38}$$

The quantity, U, in Eq. (8.36) comes from the coupling of the second derivatives, ψ_{ij}, to the function ψ. It is given by

$$U = \frac{1}{2} \sum_{i=1,j=1}^{i=4,j=4} \left[\mathbf{B}^{-1}\right]_{ij} t_i \, t_j \tag{8.39}$$

where

$$\vec{t} = (\alpha, \beta_{11}, \beta_{22}, \beta_{33}) \tag{8.40}$$

and $\mathbf{B}^{-1}$ is the inverse of the symmetric matrix whose elements are given by

$$B_{11} = \frac{1}{L^3} \sum_{\vec{q}} \frac{1}{G(\vec{q})} \tag{8.41}$$

$$B_{1\ell} = \frac{1}{L^3} \sum_{\vec{q}} \frac{q_\ell^2}{G(\vec{q})} , \quad \ell = 2, 3, 4 \tag{8.42}$$

$$B_{kk} = \frac{1}{L^3} \sum_{\vec{q}} \frac{q_k^4}{G(\vec{q})} , \quad k = 2, 3, 4 \tag{8.43}$$

and

$$B_{k\ell} = \sigma_{k\ell}^2 , \quad k \neq \ell, \quad k, \ell \neq 1 \tag{8.44}$$

The matrix $\mathbf{B}$ is somewhat simplified by the symmetry of the system which dictates that $G(\vec{q})$ is isotropic so that integrals of odd powers of q_x, q_y, q_z vanish.

Now that the probability distribution of the local gradients and second derivatives of the field ψ is known, the curvatures can be evaluated. This involves integrals of the type

$$\langle A|\vec{v}| \rangle = \tilde{P} \prod_{ij} \int d\alpha \, d\vec{v} \, d\beta_{ij} \, P\left[\alpha, \vec{v}, \beta_{ij}\right] A[\psi] \, |\vec{v}| \tag{8.45}$$

where $A[\psi]$ represents the terms in ψ and its derivatives that enter the expressions for the curvature. The term $|\vec{v}|$ comes from the surface average and $\tilde{P} \sim 1/\langle|\vec{v}|\rangle$ is a normalization that divides this expression by the average surface area. The curvature expressions contain terms such as $|\nabla\psi|$ that can be obtained from integrals performed in spherical coordinates.

The symmetry of the problem (for $\alpha = 0$) implies that $\langle H \rangle = 0$; the mean curvature of this "random" surface with Gaussian correlations vanishes. (This can be verified by an explicit calculation.) The averages of the Gaussian and mean-square curvatures are

$$\langle K \rangle = \tfrac{1}{2}\sigma_v^2 \left(\alpha^2 - 1\right) \tag{8.46}$$

$$\langle H^2 \rangle = \langle K \rangle + \frac{1}{9\sigma_v^2} [\text{Tr}[\mathbf{B}] - B_{11}] \tag{8.47}$$

where $\text{Tr}[\mathbf{B}]$ is the trace of the matrix $\mathbf{B}$ defined above.

For the symmetric case of $\alpha = 0$ (corresponding, for example, to equal water and oil volumes in a microemulsion), the Gaussian curvature is negative, corresponding to a structure that is saddle-shaped on average. On the other hand, values of $\alpha^2 > 1$ imply positive Gaussian curvature that is the case for disconnected droplets, as one would expect when, for example, the volume of water is much larger than the oil volume. The parameter α is obtained by ensuring that the volume fraction of water, ϕ_w, is given by the sum of the probabilities to find the field ψ with negative values

$$\phi_w = \int_{-\infty}^{\alpha} d\alpha' P[\psi = \alpha'] \tag{8.48}$$

The model, thus predicts a change in topology as the volume fraction is varied. The average values of the curvatures can be used to compute the bending energy of this "random" surface as a function of the correlation function, $G(\vec{q})$. One can determine $G(\vec{q})$ variationally, by calculating both the entropy and the bending energy of the system and minimizing with respect to $G(\vec{q})$ as was done in the discussion of the roughening transition in Chapter 3. (In Ref. 28, it is shown that there are additional constraints that have to be considered to conserve the amounts of water, oil, and surfactant and to avoid overcounting the entropy.) One can therefore determine consistently both the structure (via the correlation function), the free energy and phase behavior of the system as a function of the concentrations[28]. For example, the structure factor for scattering from either the oil or water domains is inversely proportional to $G(\vec{q})$ and has the form: $\left[q^4 + bq^2 + c\right]^{-1}$ where b and c are calculated functions of the concentrations and bending modulus. There is large range of concentrations where $b < 0$ and the structure factor shows the characteristic peak observed in microemulsions with a well-defined domain size.   ■

8.6 PROBLEMS

1. Growth of Cylindrical Micelles

Discuss the effect that interactions between micelles could have on their growth as a function of the volume fraction of amphiphile. For fixed number of amphiphiles in water would the addition of salt (which screens polar-head repulsions) result in longer or shorter cylinders in equilibrium?

2. Interfacial Tension at Emulsification Failure

Calculate the interfacial tension between a dispersion of water-in-oil microemulsion droplets that coexist with a phase of excess water as a function of the bending modulus and spontaneous curvature of the drops. Use the fact that at the interface there is a monolayer of surfactant that is constrained to be flat, while the spontaneous curvature implies that the lowest energy state is a curved interface.

3. Attractive Interactions and Microemulsions

Consider a water-in-oil microemulsion where there is spontaneous curvature, $c_0 = R_0^{-1}$, and the system is well described as a nearly monodisperse collection of spherical droplets of radius R. If (see the discussion in Chapter 7 on attractions in colloids) these droplets are large enough, say $R > R^*$, the van der Waals interactions between them should lead to phase separation into coexisting phases of high and low densities of drops. Sketch how this phase diagram might look as a function of the water and surfactant concentrations. Discuss what might happen in the case where $R^* < R_0$ so that the system first showed liquid/gas type phase separation and only at larger radii reached emulsification failure. What happens in the opposite case?

4. Disklike Micelles

Consider an ensemble of disk-like micelles composed of a single surfactant species. By analogy with the calculation of cylindrical micelles, find the probability distribution for finding a disk of a given size as a function of the surfactant volume fraction. Contrast the probability distribution for disks with that of cylindrical micelles (where a broad distribution of sizes exists) and comment on the reason for the difference.

Could the presence of another surfactant species stabilize a phase of large, but not infinitely large, disks?

5. Bending Constant Renormalization and Vesicle Distribution

Using the expression for the renormalized bending constant discussed in Chapter 6 in the formula for the distribution of equilibrium vesicles derived in this chapter, show how the vesicle distribution is modified by the softening of the bending modulus.

8.7   REFERENCES

1. A unified collection of articles describing the state of the field can be found in *Micelles, Membranes, Microemulsions, and Monolayers*, eds. W. M. Gelbart, A. Ben-Shaul, and D. Roux (Springer-Verlag, New York, 1994).

2. J. Israelachvili, D. J. Mitchell, and B. W. Ninham, *J. Chem. Soc. Faraday Trans. I* **72**, 1525 (1976)

3. E. W. Kaler, A. K. Murthy, B. E. Rodriguez, and J. A. N. Zasadzinski, *Science* **245**, 1371 (1989).

4. N. E. Gabriel and M. F. Roberts, *Biochemistry* **23**, 4011 (1984); W. R. Hargreaves and D. W. Deamer, *Biochemistry* **17**, 3759 (1978).

5. D. D. Miller, J. R. Bellare, T. Kaneko, and D. F. Evans, *Langmuir* **4**, 1363 (1988).

6. S. A. Safran, P. Pincus, and D. Andelman, *Science* **248**, 354 (1990); S. A. Safran, P. Pincus, D. Andelman, and F. MacKintosh, *Phys. Rev. A* **43**, 1071 (1991).

7. Z. G. Wang, *Macromolecules* **25**, 3702 (1992).

8. W. Helfrich, *J. Phys. (France)* **47**, 321 (1986), and D. C. Morse and S. T. Milner, *Europhys. Lett.* **26**, 565 (1994).

9. M. Kahlweit, R. Strey, P. Firman, and D. Haase, *Langmuir* **1**, 281 (1985); *Ang. Chem. Int. Ed. Engl.* **24**, 654 (1985); *J. Phys. Chem.* **91**, 1553 (1987).

10. For a general survey see (a) *Surfactants in Solution*, eds. K. Mittal and B. Lindman (Plenum, New York, 1984), and ibid. 1987; (b) *Physics of Complex and Supermolecular Fluids*, eds. S. A. Safran and N. A. Clark (Wiley, New York, 1987).

11. For a survey of the physics of amphiphilic systems, see *Physics of Amphiphilic Layers*, eds. J. Meunier, D. Langevin, and N. Boccara, (Springer-Verlag, New York, 1987).

12. J. Meunier, *J. Phys. Lett. (France)* **46**, 1005 (1985).

13. Reviews of globular aggregates can be found in *Micellar Solutions and Microemulsions*, eds. S. Chen and R. Rajagopalan (Springer-Verlag, New York, 1990).

14. Colloidal aspects of microemulsions are discussed in *Structure and Dynamics of Strongly Interacting Colloids and Supramolecular Aggregates in Solution*, eds. S. H. Chen, J. S. Huang, and P. Tartaglia, NATO ASI series, Volume 369 (Kluwer, Boston, 1991).

15. Y. Talmon and S. Prager, *J. Chem. Phys.* **69**, 2984 (1978).

16. P. G. de Gennes and C. Taupin, *J. Phys. Chem.* **86**, 2294 (1982).

17. B. Widom, *J. Chem. Phys.* **81**, 1030 (1984).

18. D. Andelman, M. Cates, D. Roux, and S. A. Safran, *J. Chem. Phys.* **87**, 7229 (1987); D. Andelman, S. A. Safran, D. Roux, and M. Cates, *Langmuir* **4**, 802 (1988).

19. L. Golubovic and T . C. Lubensky, *Phys. Rev. A* **41**, 43 (1990).

20. D. Huse and S. Leibler, *J. Phys. (France)* **49**, 605 (1988).

21. See the review by K. Dawson in Ref. 14; S. Alexander, *J. Phys. Lett. (France)* **39**, 1 (1978); B. Widom, *J. Chem. Phys.* **84**, 6943 (1986); G. Gompper and M. Schick, *Phys. Rev. Lett.* **62**, 1647 (1989).

22. S. A. Safran, *J. Chem. Phys.* **78**, 2073 (1983); *Phys. Rev. A* **43**, 2903 (1991).

23. S. T . Milner and S. A. Safran, *Phys. Rev. A* **36**, 4371 (1987).

24. B. Farago, D. Richter, J. Huang, S. Safran, and S. T. Milner, *Phys. Rev. Lett.* **65**, 3348 (1990).

25. S. A. Safran, L. A. Turkevich, and P. A. Pincus, *J. Phys. Lett. (France)* **45**, L69 (1984).

26. G. Porte, J. Appell, Y. Poggi, *J. Phys. Chem.* **84**, 3105 (1980); *J. Phys. Lett. (France)* **44**, L689 (1983); S. J. Candau, E. Hirsch, and R. Zana, *J.*

Colloid Int. Sci. **105**, 521 (1985). A recent review can be found in M. E. Cates and S. J. Candau, *J. Phys. Conden. Mat.* **2**, 6869 (1990).

27. M. E. Cates *et al.*, *Europhys. Letts.* **5**, 733 (1988).

28. P. Pieruschka and S. A. Safran, *Europhys. Lett.* **22**, 625 (1993).

29. N. F. Berk, *Phys. Rev. Lett.* **58**, 2718 (1987); M. Teubner, *Europhys. Lett.* **14**, 403 (1991).

30. G. Porte *et al.*, *J. Phys. II (France)* **1**, 1101 (1991).

31. H. Wennerstrom and U. Olsson, *Langmuir* **9**, 365 (1993).

32. L. E. Scriven, *Nature* **263**, 123 (1976).

Index

Printed in the United States
by Baker & Taylor Publisher Services

Printed in the United States
by Baker & Taylor Publisher Services